Habitats of the World - Biodiversity and Threats

Edited by Carmelo Maria Musarella, Ana Cano Ortiz and Ricardo Quinto Canas

Published in London, United Kingdom

IntechOpen

Supporting open minds since 2005

Habitats of the World - Biodiversity and Threats
http://dx.doi.org/10.5772/intechopen.81140
Edited by Carmelo Maria Musarella, Ana Cano Ortiz and Ricardo Quinto Canas

Contributors
Mohana S. Thota, Rasoul Karamiani, Nasrullah Rastegar-Pouyani, Eskandar Rastegar-Pouyani, Maria Teresa Capucchio, Elena Colombino, Martina Tarantola, Davide Biagini, Loris Giovanni Alborali, Antonio Marco Maisano, Federico Scali, Federica Raspa, Emanuela Valle, Ilaria Biasato, Achille Schiavone, Cristian Salogni, Valentina Bar, Claudia Gili, Franco Guarda, Koen Olthuis, Pierre-Baptiste Tartas, Chris Zevenbergen, Kasirajan Mahalingam, Antonio J. Mendoza-Fernández, Juan F. Mota, Francisco J. Pérez-García, María E. Merlo, Fabián Martínez-Hernández, Esteban Salmerón-Sánchez, Agustín Lahora, Anamaria Roman, Tudor-Mihai Ursu, Irina Onțel, Oliviu Grigore Pop, Sretco Milanovici, Teodor Marușca, Alexandru Sin-Schneider, Carmen Adriana Gheorghe, Sorin Avram, József Pál Frink, Sorina Fărcaș, Carmelo Maria Musarella, Ana Cano Ortiz, Ricardo Quinto Canas, Aroloye O. Numbere

First published in London, United Kingdom, 2020 by IntechOpen
IntechOpen is the global imprint of INTECHOPEN LIMITED, registered in England and Wales, registration number: 11086078, 7th floor, 10 Lower Thames Street, London, EC3R 6AF, United Kingdom
Printed in Croatia

British Library Cataloguing-in-Publication Data
A catalogue record for this book is available from the British Library

Additional hard and PDF copies can be obtained from orders@intechopen.com

Habitats of the World - Biodiversity and Threats
Edited by Carmelo Maria Musarella, Ana Cano Ortiz and Ricardo Quinto Canas
p. cm.
Print ISBN 978-1-78984-486-3
Online ISBN 978-1-78984-487-0
eBook (PDF) ISBN 978-1-83968-008-3

For EU product safety concerns:
IN TECH d.o.o., Prolaz Marije Krucifikse Kozulić 3, 51000 Rijeka, Croatia, info@intechopen.com or visit our website at intechopen.com.

Meet the editors

Carmelo Maria Musarella, PhD (Reggio Calabria, Italy – 23/01/1975) is a biologist, specializing in plant biology. He studied and worked in several European Universities: Messina, Catania, Reggio Calabria, Rome (Italy), Valencia, Jaén, Almeria (Spain), and Evora (Portugal). He was the Adjunct Professor of Plant Biology at the "Mediterranea" University of Reggio Calabria (Italy). His research topics are: floristic, vegetation, habitat, biogeography, taxonomy, ethnobotany, endemisms, alien species, and biodiversity conservation. He has authored many research articles published in indexed journals and books. He has been the guest editor for Plant Biosystems and a referee for this same journal and others. He is a member of the permanent scientific committee of International Seminar on "Biodiversity Conservation and Management", which includes several European universities. He has participated in several international and national congresses, seminars, workshops, and presentations of oral communications and posters.

Ana Cano Ortiz holds a PhD in Botany from the University of Jaén, Spain. She has worked in private enterprise, university, and secondary education. She was the co-director of four doctoral theses. Her fundamental line of research is related to botanical bioindicators. She has worked in Spain, Italy, Portugal, and Central America. She has written more than one hundred works published in various national and international journals, as well as books and book chapters; and has presented a hundred papers/communications to national and international congresses.

Ricardo Quinto Canas, Phd in Analysis and Management of Ecosystems, is currently an Invited Assistant Professor in the Faculty of Sciences and Technology at the University of Algarve, Portugal, and member of the Centre of Marine Sciences (CCMAR), University of Algarve. He is also the Head of Division of Environmental Impact Assessment - Algarve Regional Coordination and Development Commission (CCDR - Algarve). His current research projects focus on Botany, Vegetation Science (Geobotany), Biogeography, Plant Ecology and Biology Conservation, aiming to support Nature Conservation. Dr. Quinto Canas has co-authored many cited journal publication, conference articles and book chapters in above-mentioned topics.

Contents

Preface

The word "habitat" can have different meanings, depending on the field of study. However, in the most common sense of the word, the first thought associated with it is related to nature. However, even "nature" can be seen in different ways, but it cannot be separated from living beings and their living conditions. For these reasons, this book includes chapters on habitats under different aspects. Obviously, the main actors that we will find in them are the living beings (plants and animals in particular) that live in different parts of the Earth, but also the chemichal and physical characteristics of their habitats.

After an introductory chapter (**Habitats of the World**), the book begins with two chapters on habitats in Europe. Habitats in the European Union are protected thanks to Council Directive 92/43/EEC (Habitat Directive), adopted for the conservation of natural habitats and of wild flora and fauna. With the first chapter, the reader will have the opportunity to read about the problems related to the intensive loss of the priority habitat for *Gymnosporia senegalensis* (Lam.) Loes. [= *Maytenus senegalensis* (Lam.) Exell] in Spain. The second chapter deals with the topic of deviation from grazing optimum in the grassland habitats within and outside the Natura 2000 Network in Romania. Moving from Europe to Africa, the current problem of habitat loss continues to be deepened through the case of the Niger Delta mangroves in Nigeria. Habitat loss and conversion can lead to the extinction of mangrove forests in this region. In Asia, in particular Iran, researchers have modeled the potential distribution areas and determined the suitable habitats in the past and their current distribution for two species of snake-eyed skinks (*A. grayanus* and *A. pannonicus*). Remaining in Asia, an important case related to the floods in Korail, the largest informal settlement in Dhaka (Bangladesh), is presented. In it, authors define a "wetslum" as a specific human habitat, consisting of a slum where a considerable correlation between flooding and living conditions exists: for this reason, specific environmental and locational management for wetslums are required. Not only water, but also wind and temperature are important factors that condition a habitat. In another chapter, a case study in the Southeast peninsular of India shows the differences in thermal and dynamical characteristics and energetics of the atmosphere between wet and dry spells of the Indian summer monsoon in this region. The book ends with a chapter on the effects that "disturbed habitats" have on domestic and wild animals, focusing on pathological modifications that can derive from these disturbed habitats.

The editor would like to express his thanks to the co-editors Ana Cano Ortiz and Ricardo Quinto Canas, the Author Service Manager Sandra Maljavac and all the authors of the chapters for their support and their contributions to the publication of this book.

Carmelo Maria Musarella, PhD
Department of Agriculture (AGRARIA),
Mediterranea University of Reggio Calabria,
Italy

Ana Cano Ortiz
University of Jaen,
Spain

Ricardo Quinto Canas
University of Algarve,
Portugal

Section 1

Introduction

Chapter 1

Introductory Chapter: Habitats of the World

Carmelo Maria Musarella, Ana Cano-Ortiz and Ricardo Quinto Canas

1. Introduction

Why a book on the habitats of the world? To answer this question, we must immediately clarify what "habitat" means. This word has a clear derivation from the Latin and is a verb: more precisely, it is the third person singular of the simple present of the verb "habito, -as, -avi, -atum, -are" which means "to live." Therefore, we could literally translate the word "habitat" with "he lives." So the habitat of an organism (or of an entire community) is the place where it lives or the place where it can be found [1].

Nowadays, we use the term "habitat" in many fields of knowledge and can take on different shades of meaning (architecture, ecology, etc.).

In this book, we want to show some type of habitats of the world. The principal aim of this book is to highlight the importance of the habitats and, especially, of their inhabitants: living beings that occupy a specific space (habitat) and play a specific role in daily life in it (ecological niche). Thanks to a diversified pool of scientists belonging to several world institutions and working on several aspects of the habitats, this book will allow understanding all of it very easily.

The most important aspect that must be understood is that, respecting the habitats, the living beings present in them are respected and that the lives of others depend on them.

2. Habitat is like a home!

Often, when I want to make someone understands the meaning of "habitat," I take the example of a house with very specific characteristics, where people live. People interact with each other and with the house. All these actions occur, maintaining an equilibrium. If in a house some alterations of the normal conditions occur, this equilibrium will vary. However, when disturbing end, the equilibrium slowly comes back as before. Therefore, we can consider our house a habitat, like others in nature.

3. Habitats of the world

How many habitats are there in the world? Many! They are not easily quantifiable. Each habitat is characterized by an ecological and biological structure and different species of living beings that give it a uniqueness compared to the others.

Figure 1.
Habitat 7140 "Transition mires and quaking bogs" in the SAC IT9350134 "Canolo Nuovo Zomaro, Zillastro" (Reggio Calabria, Italy)—(Ph. C.M. Musarella).

The European Union has launched two important Community Directives for the protection of nature. With the "Habitats" Directive 92/43/CEE, the objective of safeguarding biodiversity through the conservation of natural habitats was proposed, as well as of wild flora and fauna in European territory of the member states to which the treaty applies [2]. The "Birds" Directive 79/409/EEC (the first EU directive on nature conservation) aims at the conservation of wild birds, aiming to protect the habitats of the species listed in Annex I and the migratory ones not listed that return regularly [3].

For the recognition of habitats and for the correct application of the directive, the "Interpretation Manual of the European Union Habitats" was produced, thanks to which it is possible to analyze and describe the extraordinary European naturalistic heritage [4]. All the member states applied these directives to preserve habitats in their territories. To guarantee the long-term maintenance of natural habitats and of species of flora and fauna threatened or rare at community level, the EU has established the Natura 2000 Network [5]. It consists of the Sites of Community Interest (SIC), which each member state has been identified in accordance with the Habitat Directive. Currently, the SCIs have been designated as Special Areas of Conservation (SACs) (**Figure 1**). Natura 2000 Network also include the Special Protection Areas (SPAs) established pursuant to Directive 2009/147/EC "Birds" concerning the conservation of wild birds.

There have been several scientific contributions by various scholars who have studied this important aspect of nature conservation [6–19]. The habitats are studied under several points of view, covering all their aspects [20–29]. We will also see some examples in the chapters of this book.

4. Conclusions

Based on the current world situation in which many habitats are at risk of extinction (not only, therefore, individual species), the knowledge of "places," in which every living being occupies an important part in the global balance, is more urgent than ever. Only an appropriate and adequate study of all the habitats of the world (including the discovery of new ones), from natural to anthropized ones, will allow us to preserve the biodiversity of our planet as long as possible. The study, however, must necessarily be accompanied by the will of man, the most destructive and invasive species in the world, to limit as much as possible the alterations it produces and to plan the future while simultaneously providing for adequate compensation measures. Only in this way will the "habitats of the world" have a future.

Acknowledgements

The content of this chapter is found on the activities carried out within the Research Project "Coordination of monitoring activities of natural and semi-natural habitats of the flora and fauna species of the Natura 2000 Network present in Calabria" by the Academic Editor of this book, Carmelo Maria Musarella.

Conflict of interest

The authors declare no conflict of interest.

Notes/thanks/other declarations

None.

Author details

Carmelo Maria Musarella[1,2]*, Ana Cano-Ortiz[2] and Ricardo Quinto Canas[3,4]

1 Department of Agraria, "Mediterranea" University of Reggio Calabria, Reggio Calabria, Italy

2 Department of Animal Biology, Plant Biology and Ecology, Jaen University, Jaén, Spain

3 Faculty of Sciences and Technology, University of Algarve, Faro, Portugal

4 Centre of Marine Sciences (CCMAR), University of Algarve, Faro, Portugal

*Address all correspondence to: carmelo.musarella@unirc.it

References

[1] Odum EP. Basic Ecology. Japan: Holt-Saunders; 1983. p. 325

[2] European Commission. The Habitats Directive [Internet]. 2016. Available from: http://ec.europa.eu/environment/nature/legislation/habitatsdirective/index_en.htm

[3] European Commission. The Birds Directive [Internet]. 2016. Available from: http://ec.europa.eu/environment/nature/legislation/birdsdirective/index_en.htm

[4] European Commission DG Environment Nature and Biodiversity. Interpretation Manual of European Habitats–EUR 27; 2007

[5] European Commission. Natura 2000 [Internet]. 2019. Available from: http://ec.europa.eu/environment/nature/natura2000/index_en.htm

[6] Rodwell JS, Morgan V, Jefferson RG, Moss D. The habitats directive in the UK: Some wider questions raised by the definition, notification and monitoring of grassland habitats. Plant Sociology. 2007;**44**(2 suppl. 1):37-47

[7] La Posta A, Duprè E, Maggiore AM, Tartaglini N. La Rete Natura 2000 in Italia: Un patrimonio di biodiversità da gestire/conservare e monitorare. Plant Sociology. 2007;**44**(2 suppl. 1):49-55

[8] Blasi C, Marignani M, Copiz R. Important plant areas e rete Natura 2000. Plant Sociology. 2007;**44**(2 suppl. 1):57-60

[9] Blasi C, Marignani M, Copiz R, Fipaldini M. Thematic Contribution to the National Strategy for the Biodiversity: Cartography of the Important Plant Areas in Italy (IPAs). Rome: Palombi & Partners s.r.l.; 2009. p. 31

[10] Cano E, Musarella CM, Cano-Ortiz A, Piñar JC, Pinto CJ. Vegetation series as a basis for habitats and species conservation: Methodological aspects. Botanique. 2016;**1**:21-26

[11] Spampinato G, Musarella CM. Conservation status assessment of habitats for a red list definition: The case study of Calabria. Botanique. 2016;**1**:181-182

[12] Musarella CM, Spampinato G, Cannavò S, Forestieri F, Cano E. Marsh environments management aimed at preserving endangered habitats: The case of the Aquila's Lake. Botanique. 2016;**1**:91-102

[13] Piñar JC, Cano-Ortiz A, Musarella CM, Pinto Gomes CJ, Spampinato G, Cano E. Rupicolous habitats of interest for conservation in the Central-Southern Iberian Peninsula. Plant Sociology. 2017;**54** (2 suppl. 1):29-42. DOI: 10.7338/pls2017542S1/03

[14] Musarella CM, Spampinato G, Cannavò S, Forestieri F, Cano E. Humid environments of the Messina Strait: Essential habitat for the biodiversity conservation. The case of Lake Aquila (Calabria-Italy). In: AA.VV. Action Plan—Study about Policy Measures for a Good Governance of the Messina Strait at EU Level. Provincia di Reggio Calabria. Interreg IVC Programme. 2014. ISBN 978-88-96187-04-3

[15] Spampinato G, Musarella CM, Cano-Ortiz A, Signorino G. Habitat, occurrence and conservation status of the Saharo-Macaronesian and southern-Mediterranean element *Fagonia cretica* L. (Zygophyllaceae) in Italy. Journal of Arid Land. 2018;**10**(1):140-151. DOI: 10.1007/s40333-017-0076-5

[16] Loidi J, Ortega M, Orrantia O. Vegetation science and the implementation of the habitat directive in Spain: Up-to-now experiences and further development to provide tools for management. Plant Sociology. 2007;**44**(2 suppl. 1):9-16

[17] Asensi A, Díez-Garretas B. Cartografía de los hábitat naturales y seminaturales en el Parque Natural del Estrecho (Cádiz, España). Estado de conservación. Plant Sociology. 2007;**44**(2 suppl. 1):17-22

[18] Costa JC, Monteiro-Henriques T, Neto C, Arsénio P, Aguiar C. The application of the habitats directive in Portugal. Plant Sociology. 2007;**44** (2 suppl. 1):23-28

[19] Neto C, Costa JC, Honrado J, Capelo J. Phytosociologic associations and Natura 2000 habitats of Portuguese coastal sand dunes. Plant Sociology. 2007;**44**(2 suppl. 1):29-35

[20] Bahar I, Jahan S, Rahman R. Distribution of earthworms at different habitats in Tangail, Bangladesh and significantly impacts on soil pH, organic carbon and nitrogen. American Journal of Life Sciences. 2015;**3**(3):238-246. DOI: 10.11648/j.ajls.20150303.26

[21] Adam BAA, Hassan MM, Abdelnour OM, Awadallah AH. Identification and classification of sand flies species and it's habitats in El-Kadaba Village, White Nile state, Sudan. International Journal of Infectious Diseases and Therapy. 2017;**2**(1):15-21. DOI: 10.11648/j.ijidt.20170201.14

[22] Asaduzzaman RA, Huque KS. Study on the on-farm post-natal nutritional status of red Chittagong cows in selected breeding habitat of Satkania Upazila of Chittagong District of Bangladesh. Animal and Veterinary Sciences. 2017;**5**(4):57-62. DOI: 10.11648/j.avs.20170504.12

[23] Spampinato G, Puglisi M, Privitera M, Minissale P, Sciandrello S, Noto D, et al. Which habitat for sicilian gypsophytes? In: Musarella CM, Spampinato G, editors. GYPWORLD: A Global Initiative to Understand Gypsum Ecosystem Ecology. II Workshop. Book of Abstracts and Field Trip Guide. ; 1-4 April 2019; Italy: Reggio Calabria. Italy: Mediterranea University of Reggio Calabria; 2019. ISBN: 978-88-99352-36-3

[24] Collado MÁ, Sol D, Bartomeus I. Bees use anthropogenic habitats despite strong natural habitat preferences. Diversity and Distributions. 2019;**00**:1-12. DOI: 10.1111/ddi.12899

[25] Cano Carmona E, Cano Ortiz A, Musarella CM. Introductory Chapter: Endemism as a Basic Element for the Conservation of Species and Habitats [Online First]. Rijeka, Croatia: IntechOpen; 2019. DOI: 10.5772/intechopen.84950. Available from: https://www.intechopen.com/online-first/introductory-chapter-endemism-as-a-basic-element-for-the-conservation-of-species-and-habitats

[26] Grigoriadis N, Panagopoulos A, Meliadis I, Spyroglou G, Stathaki S. Habitat and hydrological–hydrochemical characteristics of the Agras wetland (northern Greece). Plant Biosystems. 2009;**143**(1):162-172. DOI: 10.1080/11263500802709772

[27] Mendoza-Fernández AJ, Martínez-Hernández F, Pérez-García FJ, Garrido-Becerra JA, Benito BM, Salmerón-Sánchez E, et al. Extreme habitat loss in a Mediterranean habitat: *Maytenus senegalensis* subsp. *europaea*. Plant Biosystems. 2015;**149**(3):503-511. DOI: 10.1080/11263504.2014.995146

[28] Liendo D, Biurrun I, Campos JA, Herrera M, Loidi J, García-Mijangos I. Invasion patterns in riparian

habitats: The role of anthropogenic pressure in temperate streams. Plant Biosystems. 2015;**149**(2):289-297. DOI: 10.1080/11263504.2013.822434

[29] del Río S, Álvarez-Esteban R, Cano E, Pinto-Gomes C, Penas A. Potential impacts of climate change on habitat suitability of *Fagus sylvatica* L. forests in Spain. Plant Biosystems. 2018;**152**(6):1205-1213. DOI: 10.1080/11263504.2018.1435572

Section 2

Plants, Animals, Humans and Climate

Chapter 2

Intensive Habitat Loss in South Spain: Arborescent Scrubs with *Ziziphus* (5220*)

Antonio J. Mendoza-Fernández, Esteban Salmerón-Sánchez, Fabián Martínez-Hernández, Francisco J. Pérez-García, Agustín Lahora, María E. Merlo and Juan F. Mota

Abstract

The habitat arborescent matorral with *Ziziphus* (5220*) was included in the Habitats Directive of the European Commission. These plant formations represent the maximum potential vegetation in a very restrictive arid environment, since it encompasses endemic, tropical, or Maghrebian floristic elements, and from other areas of the ancient Thetis Sea. In fact, the version of this community with *Gymnosporia senegalensis* (Lam.) Loes. [=*Maytenus senegalensis* (Lam.) Exell] constitutes extraordinarily singular flora formations in the Iberian southeast. These are unique communities in Europe and ecologically extremely valuable and, however, have been included among the Europe's most endangered habitats. The vast economic development experienced in South Spain based on the remarkable transformation of traditional farming patterns into a highly profitable agriculture that uses industrial production methods and the groundwater intensively (agriculture intensification and land-use change), in addition to urbanization without sustainable land planning, determines that European *G. senegalensis* populations are seriously threatened by severe habitat destruction and fragmentation.

Keywords: habitat fragmentation, Mediterranean Basin, priority coastal habitat, semiarid ecosystems, *Ziziphus lotus*

1. Introduction

Mediterranean arborescent scrubs with *Ziziphus* Mill. species have been coded as habitat 5220* (arborescent matorral with *Ziziphus*) and included in the Habitats Directive of the European Commission since 1992 [1], which lists Europe's most endangered and vulnerable habitats. These plant communities, recognized in the Iberian Southeast, Cyprus, Sicily, and surrounding islands, are composed mainly by pre-desert deciduous scrub with *Ziziphus lotus* (L.) Lam. or *Gymnosporia senegalensis* (Lam.) Loes. [=*Maytenus senegalensis* (Lam.) Exell; =*Maytenus senegalensis* subsp. *europaeus* (Boiss.) Rivas Mart. ex Güemes & M.B. Crespo] and smaller specimens of *Periploca laevigata* Aiton subsp. *angustifolia* (Labill.) Markgraf, *Lycium intricatum* Boiss., *Asparagus horridus* L., *A. albus* L., *Withania frutescens* (Sims) Sweet, etc. The largest patches of these communities are distributed in the arid Iberian southeast under a xerophytic thermomediterranean bioclimate and correspond to the mature phase or climax of the climatophilous and edapho-xero-psammophilous vegetation series [2–4].

The Mediterranean Basin, and specifically the eastern region of Andalucia (Spain) and some islands as Sicily (Italy), accumulates a group of environmental conditions which result in the existence of these variety of habitats that have given shelter to paleoendemic species and favored specialization processes [5, 6]. This habitat, which is represented by the communities of arborescent scrub with *Ziziphus*, forms the type of vegetation that can produce the maximum of biomass in relation to the prevailing climatic conditions. These conditions include an arid and warm summer, typical of the Mediterranean climate, with low or no water availability for plants. In addition, the arrangement of this type of vegetation in hemispherical clusters is very impressive from the landscape point of view, and it has elicited various interpretations about the community dynamics [7].

2. Diagnosis

Typical scrub formations of the pre-desert climate are characterized by the presence of thorny, intricate species with leaves of small size and that are often deciduous. These types of vegetation are controlled by climatological factors such as the absence of frost, the water deficit during the dry season (high temperatures and absence of precipitation), mild annual average temperatures, and high solar radiation throughout the year [8]. The cases with greater development correspond to communities characterized by several strata of shrubs, bushes, and herbaceous species, dominated by shrubs up to 3 m high, thorny and impenetrable, which are often aggregated forming islands of vegetation (**Figure 1**). Taxa of tropical-subtropical origin or relicts of past climatic conditions, such as *Z. lotus*, *G. senegalensis*, *P. laevigata* subsp. *angustifolia*, etc., dominate them.

They develop below 300 m elevation, in semiarid and frost-free environments, on various types of substrates, although with preference for limestone, occupying depressions, riverbeds, and sporadic water flow zones, where the roots of these large shrubs could get water [9, 10].

Figure 1.
General view of the habitat 5220 and zoom on Z. lotus and G. senegalensis.*

These plant communities are very interesting for the surrounding fauna and flora, since they can create in their interior a microenvironment that contrasts with the dry and torrid conditions of the external environment, providing refuge and food to reptiles, rodents, and birds, among other groups, as well as favorable nursing processes for a number of plant species [11]. This nursing effect could be due the protection that larger plants provide against browsing of livestock or the favorable microclimatic and edaphic conditions that they promote, as it has been documented to happen in other plant communities of arid and semiarid ecosystems [11–13].

3. Distribution

This habitat is distributed through the Mediterranean Basin. In the European context, it has been inventoried in 55 Natura 2000 sites from three countries, where there are populations of species that characterize it [2]. In Cyprus, it covers an area larger than 113 ha distributed in 11 natural areas, and worthy of note is the presence of this habitat in Italy [14], which appears very locally (1.56 ha) and exclusively in three areas of Sicily, and in some surrounding smaller islands. On the contrary, in Spain, the habitat is more widely distributed and occupies more than 12,900 ha (**Table 1**). It presents in Andalucia, mainly to the south and east of the province of Almeria [15], and in more or less specific zones of the coast of the provinces of Granada and Malaga [4]. In addition, this habitat occupies the southern part of the Region of Murcia and the Valencian Community, in locations clearly exposed to marine influence and other inland territories (**Figure 2**) [16].

Site code	Country	Natura 2000 site	Cover [ha]
CY2000002	Cyprus	Alykos Potamos—Agios Sozomenos	2.78
CY2000003	Cyprus	Periochi Mitserou—Agrokipias	4.65
CY2000006	Cyprus	Dasos Pafou	0.36
CY2000008	Cyprus	Koilada Kedron—Kampos	91.29
CY2000010	Cyprus	Koilada Potamou Maroullenas	0.36
CY2000011	Cyprus	Potamos Peristeronas	0.19
CY3000002	Cyprus	Spa Kavo Gkreko	0.01
CY3000005	Cyprus	Kavo Gkreko	9.38
CY4000013	Cyprus	Faros Kato Pafou	0.43
CY6000003	Cyprus	Periochi Lympion—Agias Annas	2.58
CY6000006	Cyprus	Ethniko Dasiko Parko Rizoelias	1.29
ES0000045	Spain	Sierra Alhamilla	111.00
ES0000046	Spain	Cabo de Gata-Níjar	4024.27
ES0000047	Spain	Desierto de Tabernas	429.00
ES0000048	Spain	Punta Entinas-Sabinar	2.62
ES0000199	Spain	Sierra de la Fausilla	46.54
ES0000200	Spain	Isla Grosa	0.15
ES0000260	Spain	Mar Menor	12.44
ES0000261	Spain	Almenara-Moreras-Cabo Cope	525.31
ES0000262	Spain	Sierras del Gigante-Pericay, Lomas del Buitre-Río Luchena y Sierra de la Torrecilla	5.82

Site code	Country	Natura 2000 site	Cover [ha]
ES0000264	Spain	La Muela-Cabo Tiñoso	329.22
ES0000461	Spain	Serres del Sud d'Alacant	86.36
ES5213023	Spain	Sierra de Callosa de Segura	6.64
ES5213026	Spain	Sierra de Orihuela	33.55
ES6110002	Spain	Karst en Yesos de Sorbas	49.00
ES6110005	Spain	Sierra de Cabrera-Bédar	49.00
ES6110006	Spain	Ramblas de Gérgal, Tabernas y Sur de Sierra Alhamilla	2547.00
ES6110007	Spain	La Serrata de Cabo de Gata	57.53
ES6110008	Spain	Sierras de Gádor y Enix	616.01
ES6110011	Spain	Sierra del Alto de Almagro	1142.50
ES6110012	Spain	Sierras Almagrera, de Los Pinos y El Aguilón	1315.31
ES6110014	Spain	Artos de El Ejido	25.01
ES6110016	Spain	Rambla de Arejos	0.82
ES6110017	Spain	Río Antas	0.06
ES6140011	Spain	Sierra de Castell de Ferro	190.29
ES6140013	Spain	Acantilados y Fondos Marinos Tesorillo-Salobreña	7.22
ES6170002	Spain	Acantilados de Maro-Cerro Gordo	57.38
ES6200001	Spain	Calblanque, Monte de las Cenizas y Peña del Águila	69.33
ES6200006	Spain	Espacios Abiertos e Islas del Mar Menor	16.09
ES6200007	Spain	Islas e Islotes del Litoral Mediterráneo	2.69
ES6200010	Spain	Cuatro Calas	1.86
ES6200011	Spain	Sierra de las Moreras	133.42
ES6200012	Spain	Calnegre	184.97
ES6200013	Spain	Cabezo Gordo	18.69
ES6200015	Spain	La Muela y Cabo Tiñoso	271.82
ES6200024	Spain	Cabezo de Roldán	61.46
ES6200025	Spain	Sierra de la Fausilla	45.08
ES6200031	Spain	Cabo Cope	13.82
ES6200035	Spain	Sierra de Almenara	427.67
ES6200040	Spain	Cabezos del Pericón	3.29
ES6200044	Spain	Sierra de los Victorias	20.54
ES6200046	Spain	Sierra de En Medio	3.91
ITA010014	Italy	Sciare di Marsala	0.10
ITA020014	Italy	Monte Pellegrino	1.44
ITA090013	Italy	Saline di Priolo	0.02
Total habitat 5220* cover			13059.56

Table 1.
Natura 2000 sites where the habitat type 5220 is registered.*

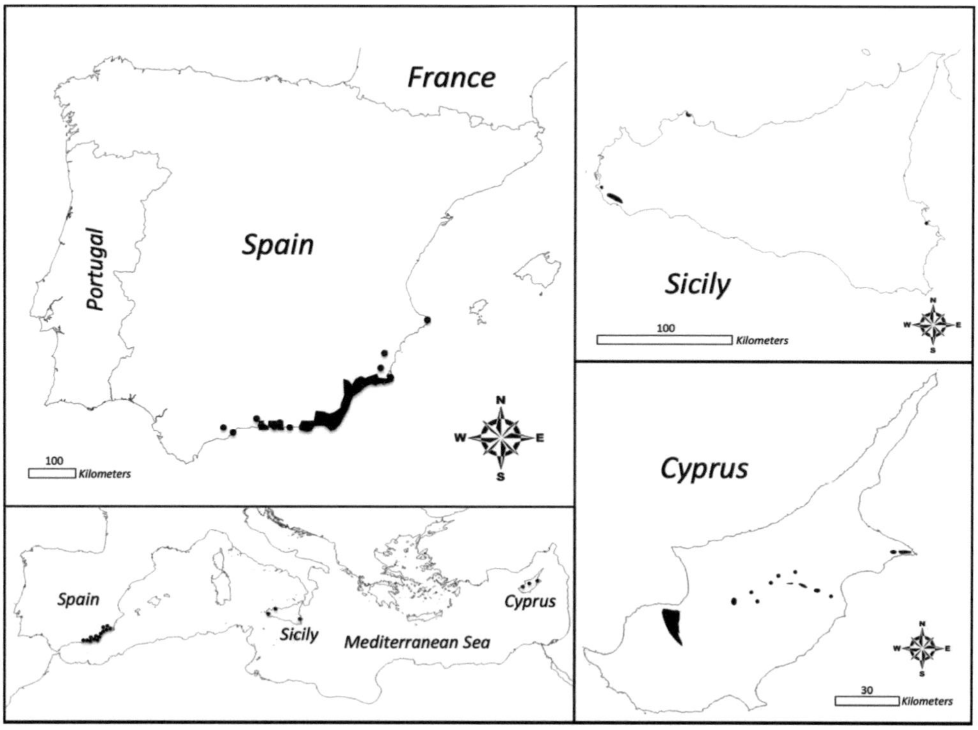

Figure 2.
Inventoried distribution of the habitat type 5220.*

4. Phytosociological associations related to habitat 5220*

The plant communities that constitute the habitat 5220* correspond to different phytosociological associations integrated in the syntaxnomic scheme of the *Quercetea ilicis* Br.-Bl. ex A. & O. Bolòs 1950 class [17–19]. Considering the geographical distribution, the general characteristics of the associations included in the habitat arborescent matorral with *Ziziphus* are listed and detailed below (**Table 2**).

In Italy, the *Z. lotus* communities are included within the alliance *Oleo sylvestris-Ceratonion siliquae*. Particularly, the association *Asparago acutifolii-Ziziphetum loti* defines the vegetation with *Z. lotus* that conserves some specimens of thorny scrubs, settled on white organogenic calcarenites, at altitudes between 5 and 75 m above sea level, in a short band of the northwest coast of Sicily. Communities nowadays relegated to disturbed or even semirural environments, near the towns and the edges of the road, are often in contact with some herbaceous or halophytic formations arranged toward localities of marine influence [20].

Otherwise, the communities of the alliance *Periplocion angustifoliae* are endemic associations with a particular phytosociological and phytogeographic interest. These communities are distributed in bioclimatic areas between the upper inframediterranean semiarid and the lower thermomediterranean dry thermotypes. *Calicotomo infestae-Rhoetum tripartitae* is an association from Sicily, composed of a xerophilous scrub dominated by *Calicotome infesta* (C.Presl) Guss. and linked to particularly arid habitats on calcareous substrates. Interesting elements of North African origin are found, such as *Rhus tripartita* (Ucria) Grande and *R. pentaphylla* (Jacq.) Desf., quite rarely in Sicily. This peculiar vegetation, now reduced to a few patches almost destroyed and fragmented, is found along the coastal strip of the

QUERCETEA ILICIS Br.-Bl. ex A. & O. Bolòs 1950
Pistacio lentisci-Rhamnetalia alaterni Rivas-Martínez 1975
Oleo sylvestris-Ceratonion siliquae Br.-Bl. ex Guinochet & Drouineau 1944
Asparago acutifolii-Ziziphetum loti Gianguzzi, Ilardi & Raimondo 1996
Periplocion angustifoliae Rivas-Martínez 1975
Calicotomo infestae-Rhoetum tripartitae Bartolo, Brullo et Marcenò 1982
Periploco angustifoliae-Euphorbietum dendroidis Brullo, Di Martino & Marcenò 1977
Ziziphetum loti Rivas Goday & Bellot 1944
Gymnosporio europaei-Ziziphetum loti F. Casas 1970
Mayteno europaei-Periplocetum angustifoliae Rivas Goday in Rivas Goday, Borja, Esteve, Galiano, Rigual & Rivas-Martínez 1960
Asparago albi-Rhamnion oleoidis Rivas Goday ex Rivas-Martínez 1975
Calicotomo intermediae-Maytenetum senegalensis Cabezudo & Pérez Latorre 2001
Oleo sylvestris-Maytenetum europaei Díez Garretas, Asensi & Rivas-Martínez 2005

Table 2.
Syntaxonomic scheme of the phytosociological associations that characterize the habitat 5220.*

southeast, in particular, in the area of Sampieri (Ragusa) in contact with formations of the *Crithmo-Limonietea* Molinier 1934 class. The vegetal community *Periploco angustifoliae-Euphorbietum dendroidis* (Sicily and surrounding islands: Pantelleria, Favignana, Levanzo, Marettimo, and Lampedusa [21–24]) characterizes a thermo-xerophilous scrub with *P. laevigata* subsp. *angustifolia* and *Euphorbia dendroides* L., of climatic sort, settled in insular coastal environments on volcanic, calcarenite, calcareous, dolomitic substrates, etc. Sometimes, the same formation can also acquire connotations of extra xericity, linked to the stoniness of the substrate in rocky or subrupicolous environments.

In Spain, several communities that are part of the habitat 5220* also integrate the same alliance (*Periplocion angustifoliae*). The *Ziziphetum loti* association defines a vegetation type composed of intricate spiny shrubs of *Z. lotus* from 1 to 3 m height, among which other species such as *Asparagus albus* and *Ballota hirsuta* (Willd.) Benth. frequently occur, as well as *Ephedra fragilis* Desf. and *Rhamnus lycioides L. subsp. oleoides* (L.) Jahand. & Maire more sporadically. The most striking aspect of the community is the mass of thorny branches, in a very apparent zigzag, that interlace with each other forming almost insurmountable barriers. During the winter, *Z. lotus* loses its leaves, while in late spring they turn in a light green shade, so the community has two highly contrasted aspects of physiognomy [9, 10, 16]. The community seems to be well settled in dry riverbeds; however the plains, which are often widely cultivated, are really its optimum. So *Z. lotus* is relegated to very strong slopes and abandoned crops.

A similar conservation concern happens in the association *Gimnosporio europaei-Ziziphetum loti*, which represents prickly scrubs up to 3 m high, dominated by *G. senegalensis* and usually accompanied by *Z. lotus* [3–5]. This is an endemic plant community of enormous uniqueness and ecological valuableness that is not found in any other part of Europe. Geographically it can be found in the southern area of the province of Almeria (Andalucia), in semiarid thermomediterranean territories with coastal influence.

The community *Mayteno europaei-Periplocetum angustifoliae* represents deciduous by drought shrub formations up to 1.5 m high, dominated by the species *P. laevigata* subsp. *angustifolia* and accompanied by sclerophyllous plants such as *Chamaerops humilis* L., *Pistacia lentiscus* L., *Rhamnus lycioides* subsp. *oleoides*, *Rubia peregrina* L., etc. It is the potential vegetation of the arid inframediterranean strip of southeastern Spain [2, 17, 25, 26]. The abundance of these formations varies from limestone soils where is high, to medium sloped silicate areas where a more open structure is usual, especially in sunny exposures. The dynamic that is related to related to the moist conditions is also remarkable, showing more typical tropical adaptations than Mediterranean ones [9, 10].

Furthermore, habitat 5220* presents in Spain a variation within the alliance *Asparago albi-Rhamnio oleidis*, where the species *Z. lotus* is not found, being the prickly species that characterizes the community is *G. senegalensis*. Two different associations have been described for this vegetation type. On the one hand, *Calicotomo intermediae-Maytenetum senegalensis* [17, 25, 26], which is a medium coverage community characterized by *G. senegalensis*, *Calicotome intermedia* C. Presl, and sporadically *Cytisus malacitanus* Boiss., developed in the thermomediterranean bioclimatic belt with a low-dry ombrotype, on calcareous soils, even of strong slopes. On the other hand, the association *Oleo sylvestris-Maytenetum europaei* represents the *G. senegalensis* thermomediterranean communities that grow in a semiarid ombrotype and are located on coastal limestone walls more or less exposed to the coastal influence or on dry riverbeds inland [3, 4].

5. Conservation of the priority habitat 5220*

This is one of the most outstanding ecosystems in Europe, whose extension of presence has been drastically reduced since the mid-twentieth century. Will the European Union be able to preserve this natural heritage? This is a priority habitat since it represents the potential natural vegetation of the territory (the expected state of mature vegetation in the absence of human intervention), that is, as the forests in other rainy territories, the Mediterranean distributions of species such as *G. senegalensis*, *Z. lotus*, and *P. laevigata* subsp. *angustifolia* indicate the maximum vegetation that the exiguous rainfall allows (**Figure 3**).

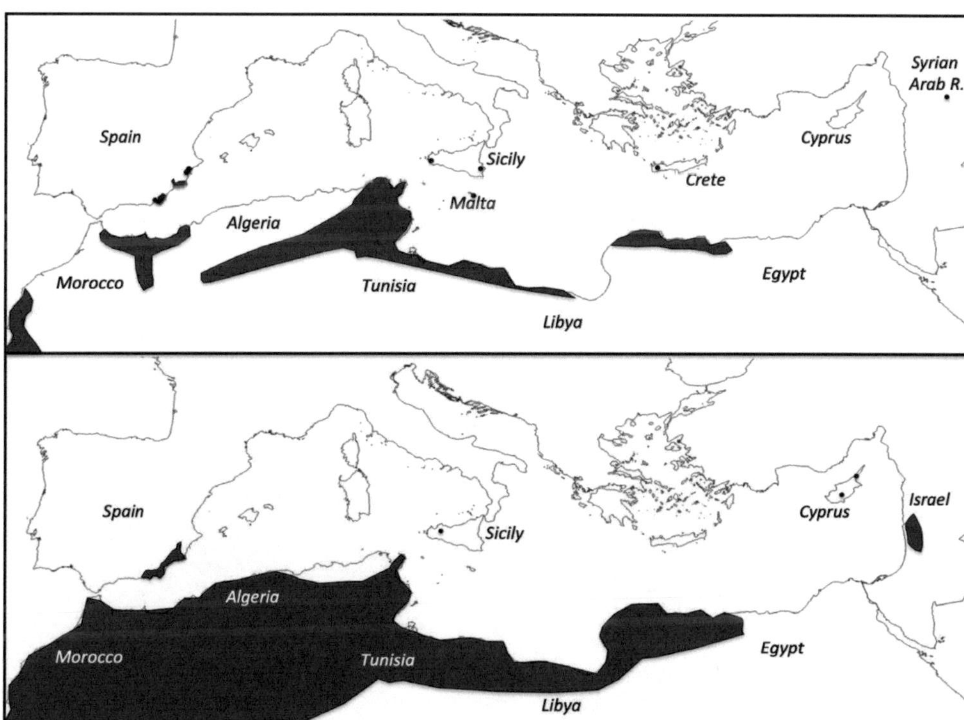

Figure 3.
Distribution of P. laevigata subsp. angustifolia (top) and Z. lotus (bottom) in the Mediterranean Basin [27, 28].

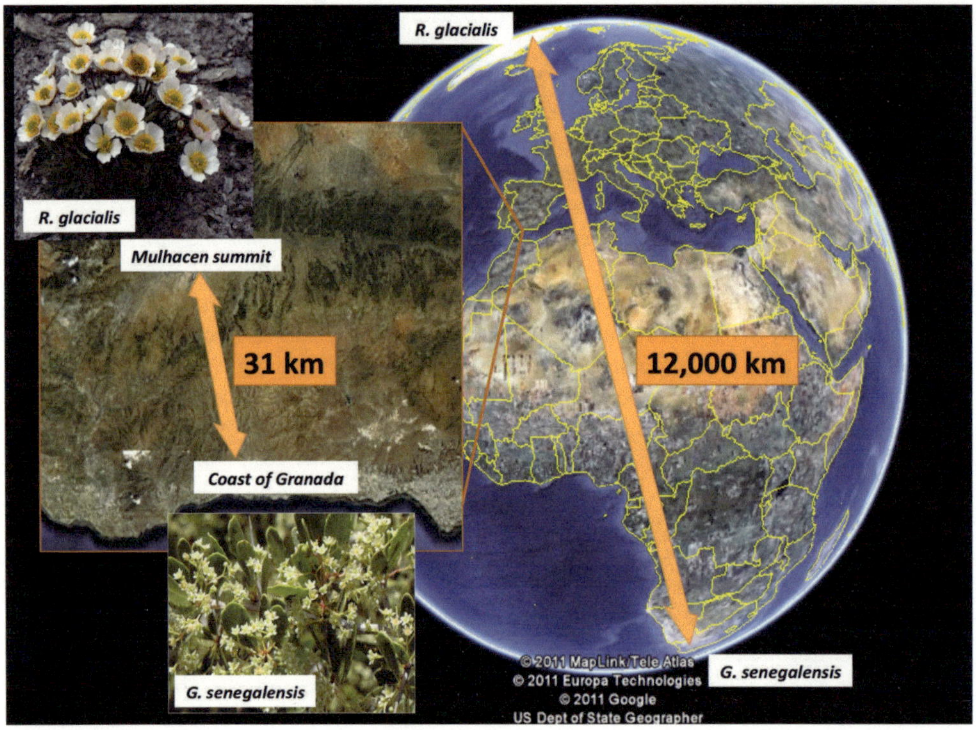

Figure 4.
Global distance between R. glacialis and G. senegalensis most remote populations and documented distance between them in southern Spain.

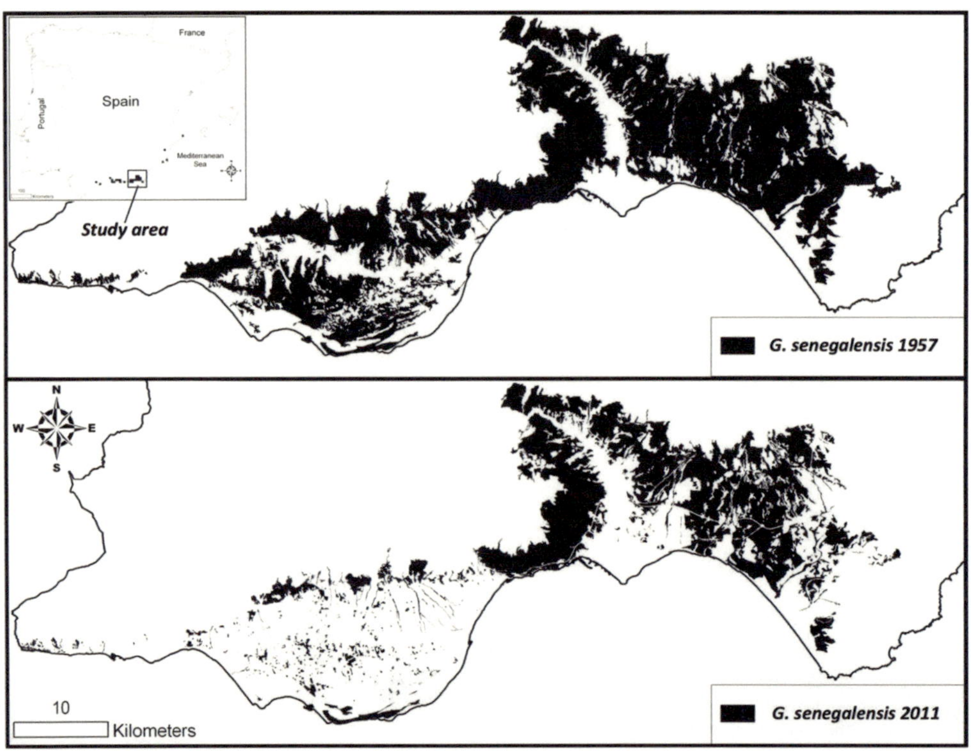

Figure 5.
Distribution area of the G. senegalensis plant formations in 1957 and 2011 in Almeria (Spain) [38].

This habitat is so peculiar that in the south of Spain, there are native flora and vegetation communities that range from *G. senegalensis* formations on the coast to alpine communities of *Ranunculus glacialis* L. The latter species, with populations on the Mulhacen summit at 3400 m,, is the vascular plant that reaches the highest northern latitude, while *G. senegalensis* reaches the coasts of South Africa (**Figure 4**). Their Spanish populations are separated by just 30 km in a straight line; and their most remote populations are separated by almost 12,000 km [29, 30].

In addition, the threat level of each species is very important [1, 31–35], but even more so is that of the communities. In fact, European *Z. lotus* habitats are seriously threatened by severe environmental destruction and fragmentation due to several risk factors such as urbanization, infrastructures, as well as agriculture intensification and land-use change [36, 38].

Some studies carried out in the southeastern of the Iberian Peninsula by combined modeling methods of environmental variables, diachronic study based in the historical photointerpretation of the area and fieldwork, showed the strong habitat regression of these communities [36–38]. Only in the province of Almeria (Andalucia, Spain), more than 26,000 ha of potential area have been lost (extension of presence) for the survival of habitat characteristic species (**Figure 5**).

6. Genetic study of the species *G. senegalensis* in Spain

Reduction in population size, so accentuated in *G. senegalensis* as a consequence of the habitat fragmentation, raises genetic barriers, since the remaining individuals are only a sample of the total number of genes present in the population [39]. Small populations may exhibit an increase in gene drift, inbreeding or outbreeding depression, and a reduction in gene flow [40–43]. The loss of genetic variability as a consequence of habitat fragmentation can have long-term evolutionary consequences and even short-term effects that involve changes at the genetic level that alter suitability and viability of the remaining populations.

Genetic structure of plant populations can be determined by a wide range of factors that interact with each other simultaneously. These factors include short- and long-term processes, such as migration, diversification, habitat fragmentation, and selection, that act at local, regional, and global range and that, when interacting with historical factors, determine geographical patterns of genetic diversity [44].

In addition to habitat loss, and due to the decrease in effective size of populations, local risks are increased by the environmental, demographic, and genetic stochasticity. Therefore, genetic variability erodes due to the random loss of alleles because of the effects of genetic drift, decreasing heterozygosity as a consequence of the increase of endogamic mating [43].

In order to clarify those questions related to the genetic structure of the populations of *G. senegalensis* that could help to establish conservation measures for this species, with the aid of the information generated by Pérez-Salmerón [44], DNA sequences have been used to detect diversity levels, in different localities through species distribution in the Iberian Peninsula (**Table 3**; **Figure 6**). With this information, we will be able to set up the basis for a next design of conservation strategies for this species.

Plant material was collected through the distribution of the species in the Iberian Peninsula (10 populations). Once in laboratory, plant material was dried in silica gel and stored at room temperature. To perform the phylogeographic analysis, DNA sequences of the ITS and trnL-trnF plastid regions were amplified from two individuals of each population, according to previous phylogenetic studies performed in *Celastraceae* family [45, 46]. Ribotype and haplotype networks obtained

Loc.	Cod.	Coord.	Alt.	H	R
El Alquián	ALQ	36°51′N 2°19'W	38	1	2
Baños de Sierra Alhamilla	BAÑ	36°57'N 2°24'W	470	1	1
Cabo de Gata	CAB	36°44'N 2°11'W	55	2	4
El Ejido	EJI	36°45'N 2°48'W	81	2	5
Melicena	MEL	36°45'N 3°14'W	17	2	2,4
Rincón de la Victoria	MSR	36°42'N 4°19'W	58	3	1,6
Cabo de la Nao	NAO	38°44'N 0°13′E	71	2	2,3
Nerja	NER	36°45'N 3°50'W	148	2	2
Velilla	MSM	36°45'N 3°38'W	116	3	2
Portman	POR	37°35'N 0°51'W	58	2	1

Table 3.
Sampled localities in the work developed by Pérez-Salmerón [44], detailing locality (Loc.), locality code (Cod.), coordinates (Coord.), altitude (Alt.), haplotypes (H), and ribotypes (R).

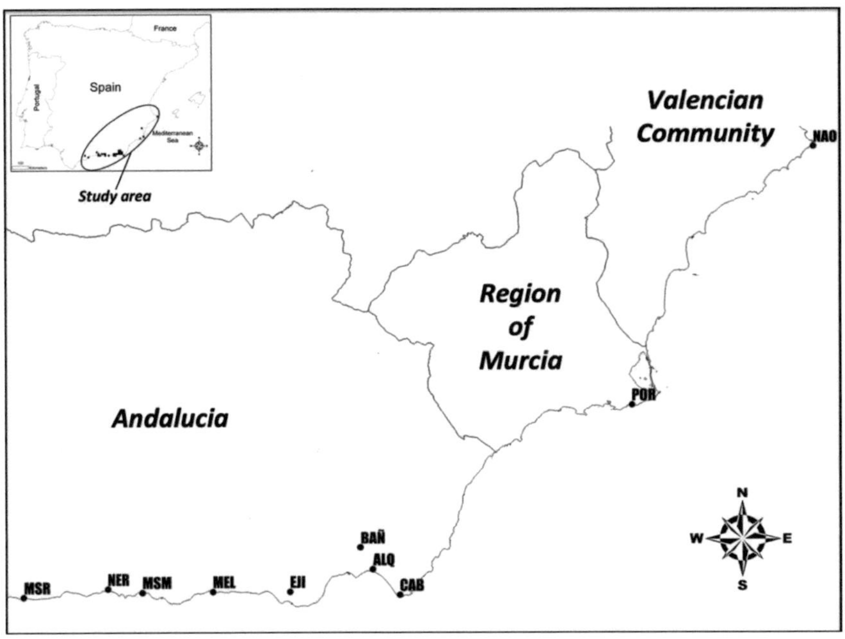

Figure 6.
Distribution of the sampled localities in the population genetic study of G. senegalensis [44].

are shown in **Figure** 7. In the case of the ribosomal sequences, among the 20 samples analyzed, it was possible to detect a total of 7 ribotypes (see **Table 3**). The highest distance between ribotypes occurred between R6 and R3, with a distance of six mutational steps. Ribotypes R2, R4, R5, and R6 were found one step away from each other. The most frequent ribotype was R2 (present throughout the distribution of the species, from NAO to MSR), followed by R1 (also of distribution in POR, MSR, and NAO) and R4 (MEL and CAB); the rest of ribotypes were present in a single locality (CAB, EJI, MSR, and NAO). With respect to plastid sequences, it was possible to detect three haplotypes. The most frequent (H2) was present in central and eastern localities (CAB, EJI, MEL, NAO, NER, and POR), whereas the remainder was present in two localities. The highest distance between the most distant haplotypes was four mutational steps.

DOI: http://dx.doi.org/10.5772/intechopen.85286

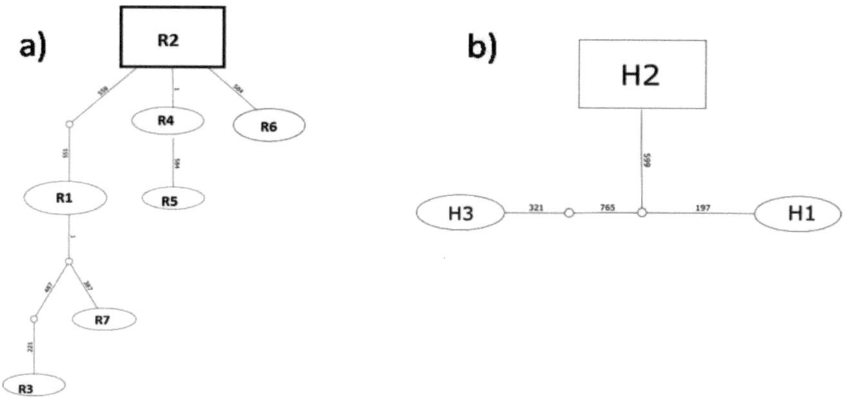

Figure 7.
Ribotype (a) and haplotype (b) networks generated by mean of the maximum parsimony algorithm as implemented in TCS software, by using nuclear (ITS1-5.8S-ITS2) and plastidial (trnL-F) sequences.

Ribotype and haplotype networks showed a low intrapopulation genetic diversity, as well as a lack of differentiation among haplotypes according to its geographical distribution. Thus, for example, R2 ribotype is present throughout the distribution of the species (ALQ, MEL, NAO, NER, MSM), while some of the haplotypes were distributed according to their distribution geographic (H1 and H3), and others not (H2). The underrepresentation of haplotypes and ribotypes in the eastern area is in line with the small area and fragmentation of these localities.

In order to maintain an in situ representation of the found lineages, it would be necessary to adopt conservation measures in at least half of the sampled populations. These populations would be BAÑ, EJI, MEL, MSR, and NAO.

7. Conclusions

This habitat is seriously threatened. The patches where this type of vegetation can be found are disappearing at a huge rate in Spain, and its presence is very scarce in Italy and Cyprus. For this reason it is consider as priority and fits in the main purpose of the Habitats Directive, which aims not only to conserve species but also entire ecosystems. The philosophy is as simple as powerful: only by protecting the community (biocenosis), all its members will be preserved indefinitely. Nowadays, to conserve implies the protection of the ecosystem, and to achieve this it is necessary to include the ecological processes and the biological components (species) that make them possible.

The lists of "Sites of Community Importance" for the conservation of the Natura 2000 Network, and the designation as "Special Areas of Conservation," have proved to be the most important initiatives for the conservation of the priority habitats. In addition, the elaboration of a checklist of characteristic species is a decisive work for the determination of these sites [47]. The management or restoration measures to ensure the favorable conservation status of the priority habitats are constituted by diverse actions implemented within the framework of LIFE+ program of the European Union (EU), among which can be mentioned, for instance, in Cyprus, the project entitled "Improving the conservation status of the priority habitat types 1520* and 5220* at the Rizoelia National Forest Park" (LIFE12 NAT/CY/000758) in which the primary aim has been to promote and enable the long-term conservation of gypsum steppes (*Gypsophiletalia*) and arborescent matorrals with *Ziziphus* in Cyprus, by quantifying and halting natural and anthropogenic pressure and threats that contribute to the long-term degradation of these habitats. In Spain, the Conhabit project 'Preservation and improvement in priority habitats on the Andalusian coast' (LIFE+13/ES/000586) is currently advancing the improvement and preservation of priority habitats found in 15 areas of the Natura 2000 Network on the Andalusian coast, and promoting social awareness of the need to protect these spaces, habitats, and species, some of which are under threat.

It is necessary to continue the elaboration of a precise cartography and the monitoring plans that help to identify the changes in the conservation status [48, 49]. The research on the plant communities should be kept open to better understand its distribution, structure, successional dynamics, and ecological requirements, especially in peripheral population patterns [44, 50]. For instance, genetic studies of *G. senegalensis* populations in Spain conclude that habitat 5220* fragmentation is associated with a progressive and drastic reduction in the size of their populations that could lead to their definitive loss. This fragmentation degree is alarming since it could have some implication with the low levels of genetic variability found (higher in the eastern region, where fragmentation and isolation are greater). These levels of genetic diversity possibly are also associated with paleoclimatic events that have contracted the area of occurrence of the species. The outcomes are worrisome considering the rate of reduction of the populations during the last decades, the adoption of measures being necessary intended for their effective protection.

Acknowledgements

This study has been made possible through the projects "Assessment, Monitoring and Applied Scientific Research for Ecological Restoration of Gypsum Mining Concessions (Majadas Viejas and Marylen) and Spreading of Results (ECORESGYP)" sponsored by the company EXPLOTACIONES RÍO DE AGUAS S.L. (TORRALBA GROUP) and "Integrated study of coastal sands vegetation (AREVEG)" sponsored by CEI·MAR.

Conflict of interest

There are no conflicts of interest.

Thanks

We want to express our gratitude to E. Pérez-Salmerón, B. J. Teruel Giménez, and J. M. López Martos, for the work produced in their final degree projects.

DOI: http://dx.doi.org/10.5772/intechopen.85286

Author details

Antonio J. Mendoza-Fernández*, Esteban Salmerón-Sánchez,
Fabián Martínez-Hernández, Francisco J. Pérez-García, Agustín Lahora,
María E. Merlo and Juan F. Mota
University of Almeria, Almeria, Spain

*Address all correspondence to: amf788@ual.es

References

[1] VVAA. Habitats Council Directive 92/43/EEC. Council of the European Union. Vol. 206. Brussels, Bélgium: DOCE; 1992. pp. 7-50

[2] VVAA. Interpretation Manual of European Union Habitats-EUR28. European Commission DG Environment; 2013. 144 p

[3] Valle F, Mapa de Series de Vegetación de Andalucía. Junta de Andalucía; 2003. 131 p

[4] Valle F, Lorite J. Modelos de gestión de la vegetación. Vol. 1-4. Consejería de Medio Ambiente. Junta de Andalucía; 2005. 512 p

[5] Valdés B. Andalusia and the Rif. Floristic links and common flora. In: Botanica Chronica. Vol. 10. 1991. pp. 117-124

[6] Blanca G. Origen de la flora de Andalucía. In: Valdés Bermejo E, editor. Introducción a la Flora de Andalucía. Junta de Andalucía. 1993. pp. 19-35

[7] Pérez-Latorre AV, Gavira O, Cabezudo B. Phenomorphology and ecomorphological characters of *Maytenus senegalensis* L. shrublands in the Iberian Peninsula: A comparison with other Mediterranean plant communities. Flora. 2010;**205**:200-210. DOI: 10.1016/j.flora.2009.04.002

[8] Tirado R. 5220 Matorrales arborescentes con Ziziphus (*). In: VVAA, editor. Bases ecológicas preliminares para la conservación de los tipos de hábitat de interés comunitario en España. Ministerio de Medio Ambiente, y Medio Rural y Marino; 2009. p. 68

[9] Valle F, Mota J, Gómez-Mercado F. Dinámica de la vegetación en el sureste de la Península Ibérica. Colloques Phytosociologiques. 1987;**15**:753-771

[10] Gorai M, Maraghni M, Neffati M. Relationship between phenological traits and water potential patterns of the wild jujube *Ziziphus lotus* (L.) lam. in southern Tunisia. Plant Ecology and Diversity. 2010;**3**:273-280. DOI: 10.1080/17550874.2010.500337

[11] Fuentes ER, Hoffmann AJ, Poiani A, Alliende MC. Vegetation change in large clearings: Patterns in the Chilean matorral. Oecologia. 1986;**68**:358-366. DOI: 10.1007/BF01036739

[12] Carrillo-Garcia Á, León De La Luz JL, Bashan Y, Bethlenfalvay GJ. Nurse plants, mycorrhizae, and plant establishment in a disturbed area of the Sonoran desert. Restoration Ecology. 2002;**7**:321-335. DOI: 10.1046/j.1526-100X.1999.72027.x

[13] Badano EI, Cavieres LA, Molina-Montenegro MA, Quiroz CL. Slope aspect influences plant association patterns in the Mediterranean matorral of Central Chile. Journal of Arid Environments. 2005;**62**:93-108. DOI: 10.1016/j.jaridenv.2004.10.012

[14] Biondi E, Blasi C, A Thematic Contribution to the National Biodiversity Strategy. Italian Interpretation Manual of the Habitats (92/43/EEC Directive). Ministry of the Environment, Land and Sea Protection, Italian Society of Botany Onlus; 2010. 18 p

[15] Mota JF, Cabello J, Cueto M, Gómez F, Giménez E, Peñas J. Datos Sobre la vegetación del Sureste de Almería. Universidad de Almería; 1997. 130 p

[16] VVAA. Manual de interpretación de hábitats Naturales y Seminaturales de la Región de Murcia; 2008. 99 p

[17] Rivas-Martínez S, Díaz González TE, Fernández-González F, Izco J, Loidi J,

Lousã M, et al. Vascular plant communities of Spain and Portugal. Itinera Geobotánica. 2002;**15**:5-432

[18] Brullo S, Marcenò C. Contributo alla conoscenza della classe *Quercetea ilicis* in Sicilia. Notulae Fitosociologicae. 1985;**19**(1):183-229

[19] eVeg. A Database Upon European Vegetations [Internet]. 2019. Available from: http://www.e-veg.net

[20] Brullo S, Cirino E, Longhitano N. Vegetazione Della Sicilia: Quadro Sintassonomico. Vol. 115. Roma: Accademia Nazionale dei Lincei; 1995. pp. 285-315

[21] Brullo S, Di Martino A, Marcenò C. La vegetazione di Pantelleria (studio fitosociologico). Istituto di Botanica Università di Catania; 1977. 111 p

[22] Bartolo G, Brullo S, Minissale P, Spampinato G. Flora e vegetazione dell'Isola di Lampedusa. Bollettino delle sedute della Accademia Gioenia di Scienze Naturali in Catania. 1988;**21**(334):119-255

[23] Di Martino A, Trapani S. Flora e Vegetazione Delle Isole di Favignana e Levanzo nell'arcipelago Delle Egadi. II Levanzo. Palermo: Lav. Ist. Giard. Col; 1966. 118 p

[24] Brullo S, Marcenò C. Osservazioni fitosociologiche sull'Isola di Marettimo (Arcipelago delle Egadi). Bollettino Delle Sedute Della Accademia Gioenia Di Scienze Naturali in Catania. 1983;**15**(320):201-228

[25] Rediam. Andalusian Environmental Information Network [Internet]. 2019. Available from: http://www.juntadeandalucia.es/medioambiente/site/rediam

[26] Díez-Garretas B, Asensi A, Rivas-Martínez S. Las comunidades de *Maytenus senegalensis* subsp. *europaeus* (*Celastraceae*) en la Península Ibérica. Lazaroa. 2005;**26**:83-92

[27] Trees and Shrubs of North Africa Platform [Internet]. 2019. Available from: http://www.northafricatrees.org

[28] Global Biodiversity Information Facility (GBIF) [Internet]. 2019. Available from: https://www.gbif.org

[29] Mota JF, Cueto M, Merlo ME, editors. Flora Amenazada de la Provincia de Almería: Una Perspectiva Desde la Biología de la Conservación. Almería: Monografías Ciencia y tecnología; 2003. 329 p

[30] Mendoza-Fernández AJ, Pérez-García FJ, Martínez-Hernández F, Salmerón-Sánchez E, Lahora A, Merlo E, et al. Red list index application for vascular flora along an altitudinal gradient. Biodiversity and Conservation. 2019;**28**(5):1029-1048. DOI: 10.1007/s10531-019-01705-y

[31] VVAA. Lista Roja de la Flora Vascular española. DGMNPF, MMA, MRM, SEBICOP; 2010. 172 p

[32] Moreno JC, editor. Lista Roja 2008 de la Flora Vascular española. DGMNPF, MMAMRM, SEBICOP; 2008. 86 p

[33] Orsenigo S, Montagnani C, Fenu G, Gargano D, Peruzzi L, Abeli T, et al. Red listing plants under full national responsibility: Extinction risk and threats in the vascular flora endemic to Italy. Biological Conservation. 2018;**224**:213-222. DOI: 10.1016/j.biocon.2018.05.030

[34] Sánchez-Gómez P, Carrión Vilches MA, Hernández González A, Guerra Montes J. Libro Rojo de la Flora Silvestre Protegida de la Región de Murcia. DGMN, CAAMA; 2002. 685 p

[35] Cabezudo B, Talavera S, Blanca G, Salazar C, Cueto M, Valdés B, et al.

editors. Lista Roja de la Flora Vascular de Andalucía. CMM, Junta de Andalucía; 2005. 126 p

[36] Mota J, Peñas J, Castro H, Cabello J, Guirado JS. Agricultural development vs. biodiversity conservation: The Mediterranean semiarid vegetation in El Ejido (Almería, southeastern Spain). Biodiversity and Conservation. 1996;**5**:1597-1617. DOI: 10.1007/BF00052118

[37] Mota JF, Peñas J, Castro H, Guirado JS, Cabello J. Los artos (*Mayteno-Ziziphetum loti*): Un hábitat en peligro de extinción. Investigación y Gestión. 1998;**2**:65-74

[38] Mendoza-Fernández AJ, Martínez-Hernández F, Pérez-García FJ, Garrido-Becerra JA, Benito BM, Salmerón-Sánchez E, et al. Extreme habitat loss in a Mediterranean habitat: *Maytenus senegalensis* subsp. *europaea*. Plant Biosystems. 2015;**149**(3):503-511. DOI: 10.1080/11263504.2014.995146

[39] Raijmann LE, Van Leeuwen R, Kerten JG, Oostermeijer B, Den Nijs HC, Menken BJ. Genetic variation and outcrossing rate in relation to population size in *Gentiana pneumonanthe* L. Conservation Biology. 1994;**8**:1014-1026. DOI: 10.1046/j.1523-1739.1994.08041014.x

[40] Aldrich P, Hamrick JL. Reproductive dominance of pasture trees in a fragmented tropical mosaic. Science. 1998;**281**:103-105. DOI: 10.1126/science.281.5373.103

[41] Nason JD, Hamrick JL. Reproductive and genetic consequences of forest fragmentation: Two case studies of Neotropical canopy trees. Journal of Heredity. 1997;**88**:264-276. DOI: 10.1093/oxfordjournals.jhered.a023104

[42] Young A, Boyle T, Brown T. The population genetic consequences of habitat fragmentation for plants. Trends in Ecology & Evolution. 1996;**11**:413-418. DOI: 10.1016/0169-5347(96)10045-8

[43] Slatkin M, Barton NH. A comparison of three indirect methods for estimating average levels of gene flow. Evolution. 1989;**43**(3):1349-1368. DOI: 10.1111/j.1558-5646.1989.tb02587.x

[44] Pérez-Salmerón E. Efectos de la fragmentación de Habitat Sobre la Diversidad genética de Maytenus Senegalensis (Lam.) Exell (Celastraceae) en el Sur de la península Ibérica. Implicaciones en Su conservación. Universidad de Almería; 2017. 39 p

[45] McKenna MJ, Simmons MP, Bacon CD, Lombardi JA. Delimitation of segregate genera of *Maytenus s.l.* (*Celastraceae*) based on molecular and morphological characters. Systematic Botany. 2011;**36**:922-932. DOI: 10.1600/036364411X604930

[46] Bacon CD, Simmons MP, Archer RH, Zhao LC, Andriantiana J. Biogeography of the Malagasy *Celastraceae*: Multiple independent origins followed by widespread dispersal of genera from Madagascar. Molecular Phylogenetics and Evolution. 2016;**94**:365-382. DOI: 10.1016/j.ympev.2015.09.013

[47] Musarella CM, Mendoza-Fernández AJ, Mota JF, Alessandrini A, Bacchetta G, Brullo S, et al. Checklist of gypsophilous vascular flora in Italy. PhytoKeys. 2018;**103**:61-82. DOI: 10.3897/phytokeys.103.25690

[48] Spampinato G, Musarella CM, Cano-Ortiz A, Signorino G. Habitat, occurrence and conservation status of the Saharo-Macaronesian and southern-Mediterranean element *Fagonia cretica* L. (*Zygophyllaceae*) in Italy. Journal of Arid Land. 2018;**10**(1):140-151. DOI: 10.1007/s40333-017-0076-5

[49] Signorino G, Cannavò S, Crisafulli A, Musarella CM, Spampinato G. *Fagonia cretica* L. In: Rossi G, Abeli T, Foggi B, Orsenigo S, Tazzari ER, Blasi C, Raimondo FM, editors. Schede per una Lista Rossa della Flora vascolare e crittogamica Italiana. Informatore Botanico Italiano. 2011;**43**(2):381-458

[50] Panuccio MR, Fazio A, Musarella CM, Mendoza-Fernández AJ, Mota JF, Spampinato G. Seed germination and antioxidant pattern in *Lavandula multifida* (*Lamiaceae*): A comparison between core and peripheral populations. Plant Biosystems. 2017;**152**(3):398-406. DOI: 10.1080/11263504.2017.1297333

Chapter 3

Deviation from Grazing Optimum in the Grassland Habitats of Romania Within and Outside the Natura 2000 Network

Anamaria Roman, Tudor-Mihai Ursu, Irina Onțel, Teodor Marușca, Oliviu Grigore Pop, Sretco Milanovici, Alexandru Sin-Schneider, Carmen Adriana Gheorghe, Sorin Avram, Sorina Fărcaș and József Pál Frink

Abstract

Grassland habitat degradation intensified in the last century worldwide and in Europe. In Romania, substantial areas of biodiverse grassland habitats that persisted due to small-scale farming are now threatened by recent land-use intensification. However, data regarding the deviation from grazing optimum, essential for management plans encompassing both socioeconomic sustainability and environment conservation, are not yet available. To fill this gap, detailed statistics of the stocking rate and its deviation from optimum were generated by spatial modeling techniques. A toolbox was developed to assess such deviations inside or outside the Natura 2000 Network of protected areas. The analysis covered an area of 33529.42 km^2, corresponding to all the Romanian permanent grasslands within the land parcel identification system. The results indicate that over half of this area is degraded, mostly from overgrazing. Less than 10% is not impacted by inadequate livestock density. Of the national grassland area, 17.34% is included within the Natura 2000 protected sites, indicating the substantial overlapping of agricultural and protection activities. For this category, the degraded area is slightly lower than at the national level (50.34% vs. 52.45%). These results can be applied for environmental conflict anticipation and optimal management of grassland habitats to achieve both socioeconomic and conservation objectives.

Keywords: vegetation, carrying capacity, grazing livestock density, grassland degradation, Natura 2000 Network, spatial modeling

1. Introduction

Grasslands are defined as herbaceous vegetation habitats with a low cover of woody vegetation, dominated mostly by grass species (family: *Poaceae*) [1, 2]. They play an important role in livestock farming but also in environmental and biodiversity conservation [3–7]. Therefore, agricultural production and nature conservation

IntechOpen

compete for the many different services that grassland habitats provide [8–10]. Although the value of grasslands, from a socioeconomic perspective and for the environmental services provided, is widely recognized, their degradation process is continuous and global [1, 11–13]. Grassland degradation generally implies a negative reduction in biodiversity, vegetation coverage, plant height, and biomass production [14–16]. Also, the deterioration of ecosystem services and functions was also included in this definition [1, 17]. The degradation process generates a series of ecological problems—loss of biodiversity, carbon sink, and water storage capacity—as well as the intensification of soil degradation and dust storms [3, 6, 18].

Worldwide, up to 50% of grasslands are affected by degradation, mainly due to human activities and climate change [12, 13, 19]. Several studies reveal that land-use changes are responsible for up to 66% of the grassland degradation, whereas the climate dynamics account for approximately 20% [13, 14, 19, 20]. At a European level, climate is the primary degradation agent in some areas of Northern and Northwestern Europe and the southern part of European Russia, but in most areas, including Eastern Europe, degradation is mainly caused by land-use issues [13, 21]. Sudden changes in land-use intensity such as overgrazing or abandonment of traditional farming practices are among the main factors identified to cause the degradation of grassland habitats [13, 22–25]. The alteration of agricultural practices (intensification or abandonment), along with the area of degraded grasslands and the associated environmental problems, shows an upward trend [26–28].

The most important policies aiming to manage and mitigate these issues that have been developed within the European Union (EU) are the Common Agricultural Policy (CAP) framework and the Environmental Directives (especially Habitats Directive 1992/43/ECC-HD and Birds Directive 2009/147/EC-BD). For grassland habitats, these policies mainly focus on agricultural production (livestock density) and, respectively, on biodiversity conservation. The CAP directives include livestock density determination according to the grassland carrying capacity, while the Habitats Directive (1992/43/ECC) implements the conservation of the habitats of community importance which were selected according to their structure (floristic diversity) and environmental ecological functions. The favorable conservation status of these habitats must be reached or maintained within all the sites which are included in the European Natura 2000 Network (N2000). The network includes a large number of protected areas (27863 sites), being acknowledged as one of the world's most effective legal instruments for biodiversity and nature conservation, with an important function in conserving Europe's natural capital. It is estimated that approximately 16% of the habitats in N2000 areas depend on a perpetuation of extensive farming practices and especially on maintaining the extensive management of grasslands [29]. In the EU-27, approximately 18% of the permanent grasslands are within the protected N2000 Network [30]. However, the effects of grazing livestock density (stocking rate) on the grassland habitats protected within N2000 sites have rarely been considered so far, particularly the context of their actual spatial overlap. Moreover, for some countries hosting very large areas of permanent grasslands in the EU (e.g., Romania), spatially detailed data at the landscape level are not yet available, although the agricultural statistics are reported at the national level by each member state. For instance, the spatial distribution of livestock in Europe was modeled using statistical downscaling of province-level livestock statistics [31], but the possible deviations from the grassland habitats' optimal livestock density (carrying capacity) are not yet assessed.

This paper aims to evaluate the degradation status of grassland habitats by modeling and mapping the grazing livestock density and the subsequent deviations from the optimal grassland carrying capacity within and outside the N2000 sites from Romania. The permanent grassland habitats from Romania are among the most extensive and diverse from the EU [23, 32]. They cover more than 45,000 km^2, which represents 8% of EU-27 permanent grassland habitats. Only the UK (17%), France (15%), Spain (15%), and Germany (9%) have a larger grassland area [32]. Also, the legally protected Sites of Community Importance (SCI) from N2000 Network which are designed for the conservation of all the habitats enlisted in the HD now cover 16.7% of the EU land surface and 19.5% of Romania's total territory. Detailed knowledge regarding the spatial patterns of grazing intensity within and outside the N2000 sites is therefore needed in order to identify the areas where the intensity of agricultural practices is divergent from optimum and particularly where these and nature conservation efforts overlap. Also, the identification of such areas may serve as a basis for land managers and agriculturists but also for the organizations involved in biodiversity conservation to better design the grazing patterns and protection measures in order to avoid grassland degradation either by intensification or abandonment. In the context of the high value of grassland habitats, both from socioeconomic and ecological reasons, this approach provides a meaningful perspective on the relationship between agricultural land values and nature conservation for all relevant stakeholders. This insight could further support policies aiming at a future conflict-free combination of agricultural production and nature conservation. Detailed statistics regarding the deviation from the optimal livestock density were generated by spatial modeling techniques in a geographic information system (GIS) environment. A GIS toolbox was developed for the spatial modeling of these deviations inside or outside of the protected areas. This can be used for the environmental conflict anticipation and subsequent management of the grassland habitats so that both socioeconomic and conservation targets are achieved.

2. Materials and methods

2.1 Study area

Romania (area 238397 km^2; capital Bucharest 44°25′57″N, 26°06′14″E) is located in Southeastern Europe, bordering on the Black Sea and the Danube, with the Southeastern Carpathian Mountains in its center (**Figure 1**). The natural landscape includes almost even proportions of mountains (31%), plains (33%), and hills (36%) that expand rather uniformly from the mountains, reaching elevations above 2500 m, to the Danube Delta, a few meters above sea level. The climate is transitional between temperate and continental. The average annual temperature goes from 11°C in the south to 8°C in the north. Annual precipitation decreases eastward and downward, averaging up to 1010 mm in some mountainous areas, 635 mm in the Transylvanian Plateau, 521 mm in Moldavia, and only 381 mm in Dobruja and close to the Black Sea.

The Corine Land Cover dataset [33] reports for Romania the following land cover classes: artificial areas (5.34%), arable land and permanent crops (39.37%), pastures mosaics (17.65%), forested land (31.68%), seminatural vegetation (2.78%), open space and barren soils (0.10%), wetlands (1.35%), and water bodies (1.69%). The General Agricultural Census in Romania, performed in 2010, indicates that the permanent grasslands cover 44940 km^2, including both grazed pastures and hay meadows, that together make up about 33% of total utilized agricultural land [32, 34]. The greatest surface covered by permanent grasslands is in the Carpathian Mountains region and in the Transylvanian Plateau, where every county has between 1000 and 3500 km^2 of grassland.

The geomorphological and climatic diversity of Romania, the geographical position at the intersection of several floristic provinces, and the extensive traditional

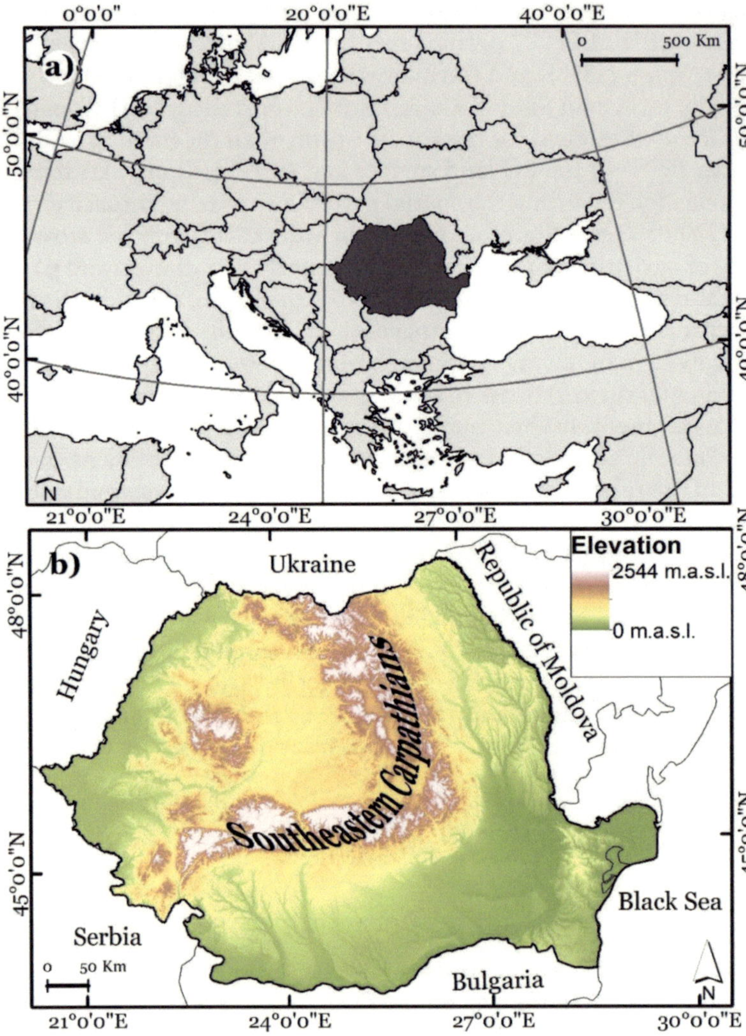

Figure 1.
The geographical position (a) and the elevation model (b) of Romania.

land use all contribute to the vegetation diversity [35, 36], reflected also in the large variety of grassland habitat types [4, 37, 38]. Most herbaceous vegetation types (except ruderal) are comprised in 15 N2000 grassland habitat types [39, 40].

The Romanian grassland habitats are diverse, including dry grasslands, mesophilous grasslands, high-mountain grasslands, and wet grasslands. The detailed description of the floristic structure specific to these vegetation types can be found in phytosociological studies [41] and the Romanian grassland inventory [42]. According to the latter source that mapped an area of 3900 km^2, the best-represented habitat types were mesophilous (39.1%, mostly *Arrhenatheretalia* vegetation order) and dry grasslands (38.2%, mostly *Festucetalia valesiacae*), followed by high-mountain grasslands (12.7%, *Nardetalia*, *Caricetalia curvulae* etc.), wet grasslands (5.35%, mostly from *Molinietalia*), and ruderal-degraded grasslands (4.2%).

2.2 Data and spatial modeling

The spatial distribution of the deviations from the optimum livestock density (DEV_{OLD}) was modeled in GIS in order to quantify and map its effect on the grassland habitat degradation status throughout the grassland habitats from Romania. The data presented in **Table 1** were geoprocessed in ModelBuilder, and a GIS toolbox was developed for analyzing the DEV_{OLD} (grazing carrying capacity). All the GIS processing and spatial analysis were performed using ArcGIS 10.5 [43].

This study encompassed all the permanent grassland polygons (GP) (33529.42 km^2) which are included in the Land Parcel Identification System from Romania [44]. The permanent grassland area is the land used to grow grasses or other herbaceous forage that is not subject to crop rotation for at least 5 years or longer [32, 34, 44]. Also, the data regarding the area (40451.91 km^2) covered by all the 435 Romanian Sites of Community Importance (ROSCI) was included in the model. The dataset with the numbers and types of livestock from each TAU was downloaded from the National Statistics Institute of Romania [34]. The total number of the different livestock types (animal heads) was recorded during the General Agricultural Census from 2010 for 3177 TAUs from 41 counties. Only the following types of grazing livestock were included in the analysis: cattle (dairy and beef), sheep, goats, horses, donkeys, and mules. Livestock numbers were converted into livestock units (also called animal units) using specific coefficients indicated in the official Romanian guidelines [45]. The livestock unit (LU) is a reference unit which facilitates the aggregation of livestock from various species and ages. One LU is the grazing equivalent of one adult dairy cow producing 3000 kg of milk annually, without additional concentrated foodstuff. According to the transformation coefficients from the national regulations [47–49], the formula and coefficients used for the conversion of the animal numbers and types in number of LUs (for each TAU) is:

$$\text{LU number} = \text{Cattle Number} + (\text{Sheep Number} \times 0.15) + (\text{Goats Number} \times 0.15) + \text{Horses Number} + (\text{Donkeys \& Mules Number} \times 0.4) \quad (1)$$

Input data	Data source
Permanent grassland polygons (GP)	Land parcel identification system from Romania [44]
The polygons of territorial administrative units (TAU)	http://geoportal.ancpi.ro/geoportal
The polygons of N2000 Sites of Community Importance (ROSCI)	http://www.mmediu.ro/articol/date-gis/434
Digital elevation model (DEM 25 m)	Digital elevation model over Europe (EU-DEM) https://www.eea.europa.eu/data-and-maps/data/copernicus-land-monitoring-service-eu-dem#tab-gis-data
Livestock numbers and types from TAUs	National Statistics Institute of Romania—The General Agricultural Census 2010 [34]
Coefficients for the transformation of different animal types in livestock units (LU)	The official regulations in Romania regarding the methodology for the evaluation of the optimal livestock density per hectare [45, 46]
Optimal livestock density for socioeconomic production (OLD_E)	The official regulations in Romania regarding the grazing management plans and grassland experts [47–49]
Optimal livestock density for biodiversity conservation (OLD_B)	The upper limit/level for the optimal livestock density recommended by various studies for biodiversity conservation in Central and Southeastern European countries [21, 50–54]

Table 1.
The input data for spatial modeling of the deviations from the optimal livestock density within and outside the N2000 sites from Romania.

The total livestock density (LD) measures the stock of animals, expressed in LU, per hectare of permanent grasslands. LD was calculated considering the total number of LUs from each TAU divided by the total area of permanent grassland of the respective TAU. Since there are no available data regarding the spatial distribution of the LD within each TAU from Romania, the LD (LU/ha) was calculated for the permanent grassland area of each TAU. Also, this approach is supported by the fact that a single grazing management plan is designed for all grassland parcels at the TAU level [45].

The difference between the current LD of a grassland and the optimum livestock density for the respective area and conditions represents the deviation from OLD. The equation for generating the deviation from optimum livestock density (DEV_{OLD}) in each grassland polygon is:

$$DEV_{OLD} = \frac{LD \times 100}{OLD} - 100 \quad (2)$$

LD is the livestock density as livestock units/hectare (LU/ha); OLD is the optimum livestock density for the grassland polygon.

The areas where the deviation from the DEV_{OLD} is at most plus or minus 10% are considered not impacted. The fragments of grassland habitats where the DEV_{OLD} is between 10.1 and 50% (plus or minus) are considered partially impacted. Those where DEV_{OLD} is over 50% (plus or minus) are considered to be subject to a major impact, and degraded because of inappropriate livestock densities [48, 49]. Impact and degradation can be caused both by overgrazing and abandonment. In the first case, the grass cover decreases and allows the expansion of ruderal species that are good competitors but have a low forage value, or, worse, the soil is stripped of vegetation, favoring erosion. In the second case, the abandonment of grassland usage (as pasture or hayfield) is also harmful, resulting in shrub invasion which decreases grassland biodiversity and finally in the establishment of forest habitats.

Two scenarios were considered for the grassland habitats included in the N2000 ROSCIs. The first one, which was applied for all the grassland habitats of Romania, employs an optimum livestock density considered suitable for the grassland habitat areas with predominant socioeconomic purpose (OLD_E). OLD_E was synthesized from the Guidelines for Elaborating the Grazing Management Plans [47] that takes into account the different ecological and production characteristics of the various grassland habitat types. As a consequence, for each of the three main altitude belts of Romania, a specific OLD (LU/ha) was assigned, as follows: 0.46 (20–200 m a.s.l.), 0.6 (201–800 m a.s.l.), and 0.9 (801–2544 m a.s.l.).

The second scenario analyzes the prospect of using a lower OLD, favoring biodiversity conservation (OLD_B), for the grasslands situated within N2000 ROSCIs, where lower intensity grazing is recommended [21, 50–54]; in the case of OLD_B, the value of 0.45 LU/ha [21] was employed, although the large range of elevations and ecological conditions from the territory of Romania might require more specific values for different grassland types and altitude belts.

3. Results and discussions

The assessment and mapping of the deviation from the grazing carrying capacity were carried out within an area of 33529.42 km^2 that corresponds to the permanent grasslands from Romania. Our results indicate that 17.34% (5814.75 km^2) of these grasslands are situated within the N2000 ROSCIs (**Figure 2**). This indicates an important overlap between domestic livestock husbandry and nature conservation within ROSCIs, both supported by the EU within the rural, regional, and environmental development policies. The grazing livestock types (cattle, sheep, goats, horses, donkeys, and mules) from each TAU, presented in **Figure 2**, depend on these grassland habitats, which are the most important resource for livestock production systems [34]. It is estimated that permanent grasslands provide at least 60% of the forage necessary for cattle and 80% for sheep [49]. The livestock density is considered to be one of the most relevant indicators of grassland degradation status, being strictly connected to both the socioeconomic factors and the ecological carrying capacity of grassland habitats. Overstocking permanent grasslands as well as understocking them until abandonment impacts them and, at high intensities, causes their degradation.

However, the time span between successive grazing events may also be very important besides LD and grazing intensity [31, 55], but it is very difficult to quantify and map at large scale for each individual grassland polygon. Modeling the livestock data reported at the TAU level for evaluating the LD and DEV_{OLD} is the best alternative for the available data, although it has the limitation of assigning the same values and status for all the grasslands within a TAU.

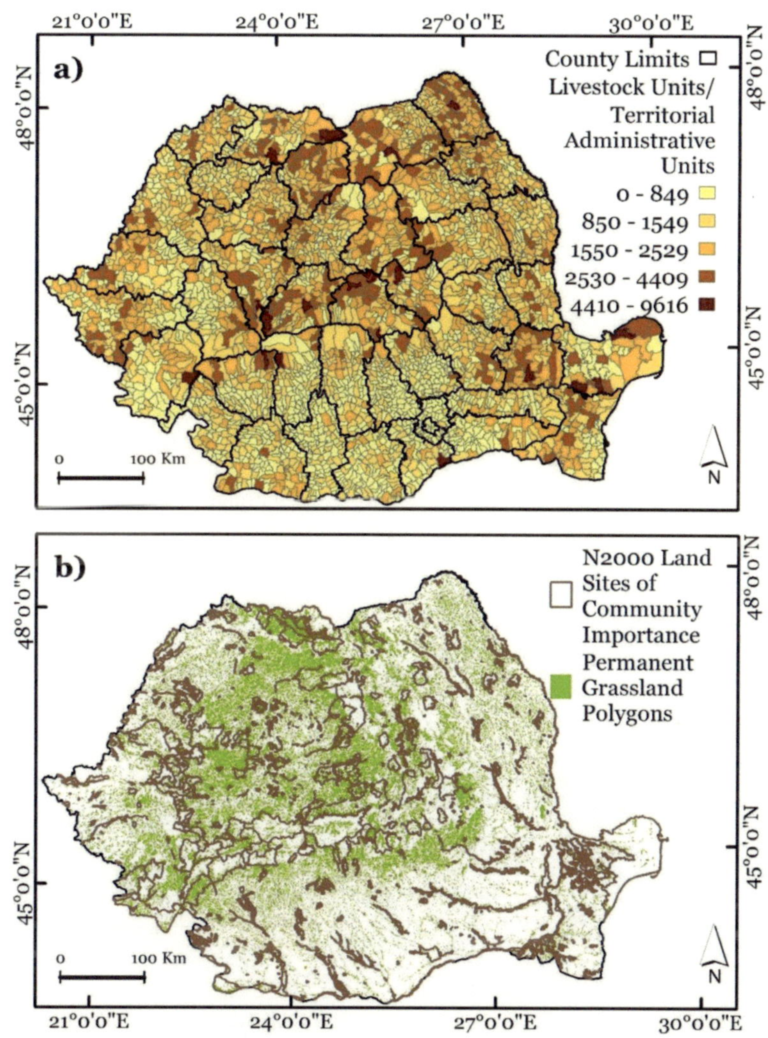

Figure 2.
Livestock distribution in the territorial administrative units from Romania (a). The permanent grassland habitats and the limits of the N2000 Sites of Community Importance from Romania (b).

The geoprocessing steps that were performed and integrated into ModelBuilder in order to identify the status of each grassland polygon regarding the DEV_{OLD} are presented in **Figure 3**. In the first stage, all the input data were processed at the national level. The current LUs and subsequently the LD (LUs/ha), OLD, and DEV_{OLD} were derived for each grassland polygon based on the OLD_E values recommended for each of the three altitude belts from Romania (**Figure 3a**). In the second stage, the resulting grasslands–DEV_{OLD_E} dataset was intersected with the limits of ROSCIs in order to analyze the status of the grassland habitats included within these protected areas (**Figure 3b**). Subsequently, in the third stage, the OLD_B value was input into the model as an alternative to OLD_E for the grassland habitats included in ROSCIs (**Figure 3c**). The developed GIS toolbox with the OLD model is flexible, allowing to easily test a different OLD or to be adapted for any similar case study. As mentioned above, the results obtained from the model are an approximation that considers the LD as having a uniform distribution throughout all the grassland habitats from each TAU. Although the situation within individual grassland parcels might be different, on average it is accurate at the TAU level, particularly taking into account the spatial and temporal dynamics of grazing, the high probability

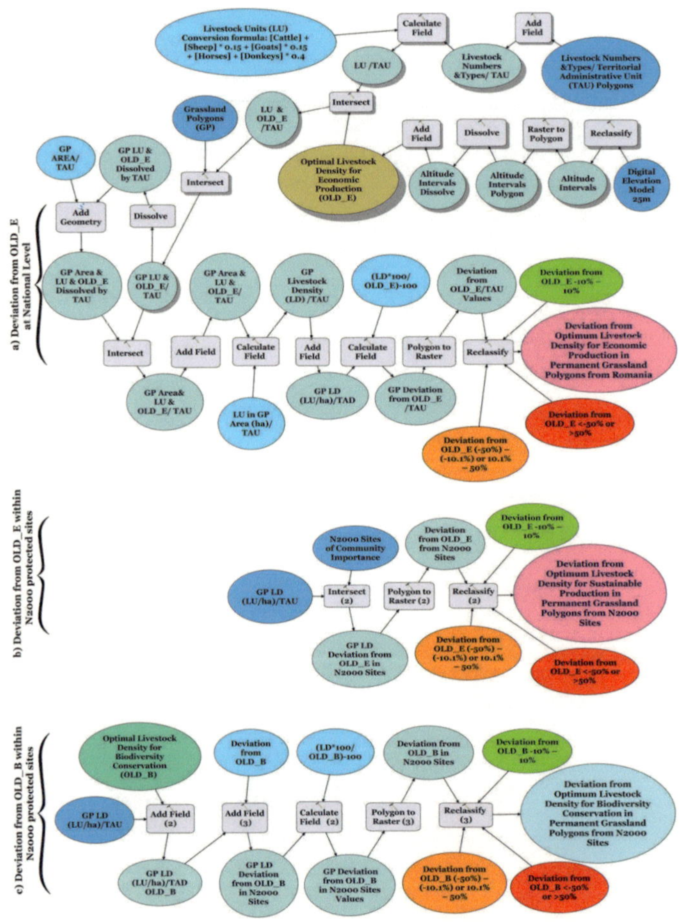

Figure 3.
The models generated for the analysis of the (a) grassland habitats at the national level, using deviations from the optimal livestock for socioeconomic production (OLD_E); (b) grassland habitats from the N2000 SCIs, using deviations from the optimal livestock for socioeconomic production (OLD_E); and (c) grassland habitats from the N2000 SCIs, using deviations from the optimal livestock for biodiversity conservation (OLD_B).

of livestock grazing within the TAU of their owners, and the grazing management plans which are designed at the TAU level.

The assembly of the input data within the OLD model and the used spatial analysis tools are presented below for the grassland habitats situated within and outside protected areas (**Figure 3**).

3.1 Case scenario 1: status of the grassland habitats with the optimal livestock density for sustainable economic production (OLD_E)

This scenario considers the values of the optimal livestock density (carrying capacity) recommended by the Romanian grassland experts for sustainable economic production of biomass [47–49]. Most grazing management studies and textbooks recommend different optimal LDs (stocking rates), but they generally tend to increase with altitude following the available plant biomass.

For the analyzed grassland polygons of Romania (33529.42 km^2), the deviation of the existing livestock density from the OLD_E (**Figure 4**) results in 52.45% of the grassland area being degraded (major impact, current LD with more than 50% over or under OLD_E). The LD was much higher than the carrying capacity (overgrazing impact) for 44.05% of the area, with 8.40% of the area being impacted by abandonment, the LD being far under the OLD. Of the 39.25% grassland habitat area that is partially impacted (10.1–50% over or under the CC), 23.94% has an LD under the optimal value, while 15.31% is moderately overgrazed. At the national level, only 8.28%

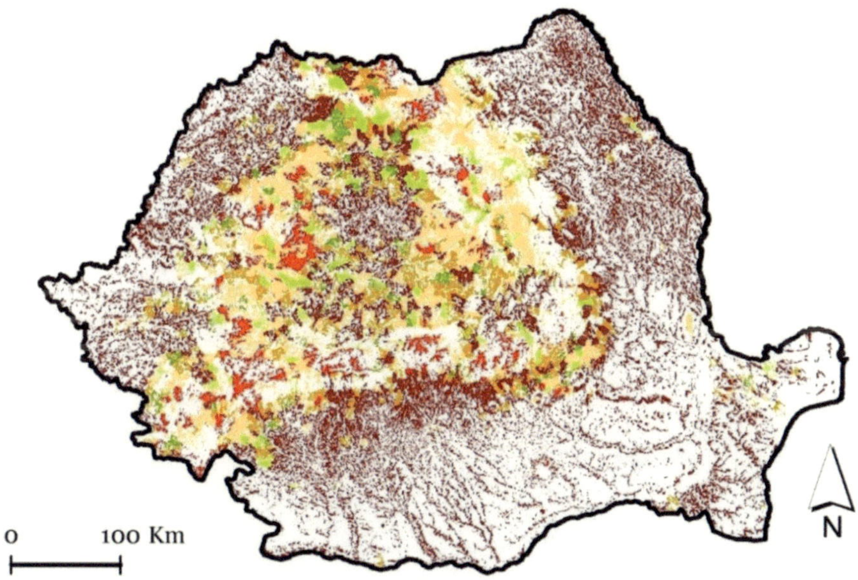

Deviation from Optimal Livestock Density for Economic Production	Grassland Habitat Status in Romania	Percent (%)	Area (km²)
-10% – 0%	No impact - Slight Abandonment	4.91	1645.93
0% – 10%	No impact - Minor Overgrazing	3.37	1131.27
-50% – -10.1%	Partial impact - Moderate Abandonment	23.94	8026.25
10.1% – 50%	Partial - Moderate Overgrazing	15.31	5132.02
< -50%	Major impact (Degraded)- Abandonment	8.40	2817.70
> 50%	Major impact (Degraded) - Overgrazing	44.05	14770.09
No Data	No Data	0.02	6.17

Figure 4.
The spatial distribution of impact and degradation at the national level caused by the deviations from grazing optimum for the socioeconomic production scenario; the deviation classes, status, their percentage, and area at the national level.

of the permanent grassland area is not impacted with regard to the carrying capacity, the LD being within the interval of 10% over to 10% under the optimal value, 3.37% being partly overgrazed, while 4.91% of this area being used slightly below the optimal intensity. Although the obtained results concerning the LD distribution and deviation from OLD are consistent with other studies that evaluate the grassland status from Romania [31, 32, 48, 56], they are only supported by the livestock statistical reports, field data regarding the habitat status not being available yet. Validation by grassland experts in the field was only performed for one TAU (Zăvoi) [48] for a model that used the same dataset (livestock units within the permanent grassland polygons of a TAU) to extract the livestock density classes and evaluate the grassland status.

When analyzing the situation of the grassland polygons that are within N2000 ROSCIs, the percentage of the degraded area is 50.33%, slightly lower than at the national level (**Figure 5**). Of this, 30.52% represents areas prone to degradation from intense overgrazing (the current LD being far over the OLD), while in 19.81% of the area degradation is caused by abandonment. The proportion of partially impacted grasslands is slightly larger within the N2000 sites than at the national level, reaching 42.99% of the total area. Most of these grasslands are undergoing moderate abandonment (28.86%), while moderate overgrazing affects an area only half as large (14.13%). A smaller proportion of the ROSCI grasslands—6.68%—are not impacted with regard to this criterion than at the national level. Of these grasslands, those experiencing minor overgrazing and the ones with slight abandonment have similar percentages, 3.22% and 3.46%, respectively.

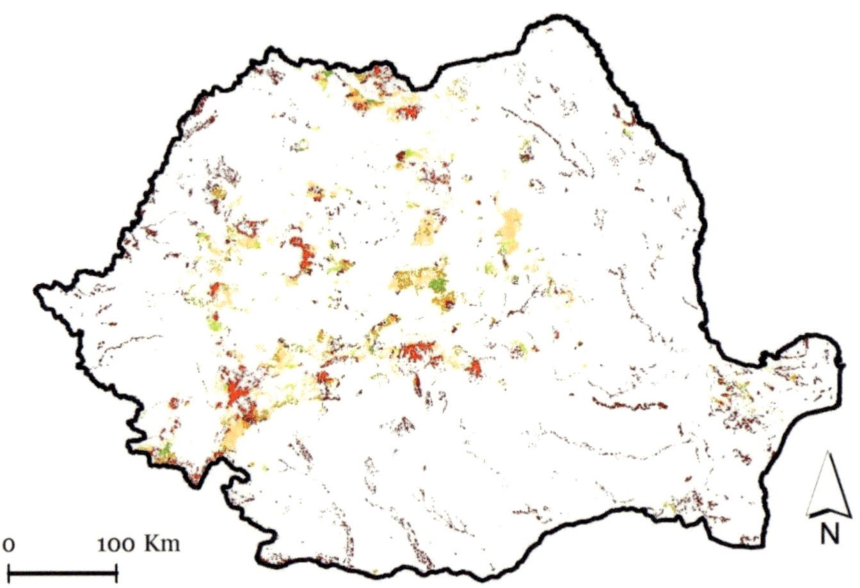

Deviation from Optimal Livestock Density for Economic Production	Grassland Habitat Status in Natura 2000 sites (SCI) from Romania	Percent (%)	Area (km²)
-10% – 0%	No impact - Slight Abandonment	3.46	200.77
0% – 10%	No impact - Minor Overgrazing	3.22	187.12
-50% – -10.1%	Partial impact - Moderate Abandonment	28.86	1675.35
10.1% – 50%	Partial - Moderate Overgrazing	14.13	820.14
<-50%	Major impact (Degraded)- Abandonment	19.81	1149.76
>50%	Major impact (Degraded) - Overgrazing	30.52	1771.86
No Data	No Data	0.01	0.32

Figure 5.
The spatial distribution of impact and degradation within the N2000 ROSCIs caused by the deviations from grazing optimum for the socioeconomic production scenario; the deviation classes, status, their percentage, and area at the national level.

3.2 Case scenario 2: status of the grassland habitats with the optimal livestock density for biodiversity conservation (OLD_B)

For the grasslands included in N2000 ROSCIs, a lower livestock density might be recommendable that has been shown to maintain and enhance biodiversity in similar contexts from Central and Southeastern Europe [21, 50–54]. For the analysis, the value of 0.45 LU/ha was tested, the recommended LD for biodiversity conservation being under 0.5 [21, 50–54]. However, the large range of elevations and ecological conditions from Romania might require more specific values for different grassland types and altitude belts if a more accurate evaluation is desired. In the perspective of this lower optimum LD (**Table 2**), an 8% increase in the proportion of degraded N2000 grasslands appears (to 58.49%), the major impact source being overgrazing, 50.46%, while abandonment contributes with only 8.03% to degradation. The percentage of partially impacted grassland areas is lower than in the previous scenario by 33.35%, while overgrazing and abandonment have almost equal importance in this case (17.91% and 15.44%, respectively). In this scenario, the percentage of the area not impacted is 8.15%, very similar to the nationwide figure for all the grasslands. Of this area, 4.41% is lightly overgrazed, while 3.74% is used slightly below the optimal intensity.

It appears that in the case of the grasslands from N2000 sites, which are important for biodiversity conservation, under the existing LD conditions, a lower optimal LD can be proposed without generating widespread conflicts between the socioeconomical activities and nature conservation. Since the major impacts include both overexploitation and abandonment, in some neighboring TAUs that experience opposite tendencies, sharing the grassland resources and a better distribution of the livestock may be a first, easier step to improve grassland degradation status (**Figure 6**). Our results regarding the areas free from overgrazing are consistent with other studies which revealed that the spatial (geographical) distribution of grazing may be as important as the LD [24, 57, 58]. This means that, beyond good local management of grazing, an optimized, larger scale of grazing management is needed as well. When viewed at a regional scale such as TAUs, where we graze may be as important as how we graze.

DEV_{OLD}	Grassland habitat status	DEV_{OLD_E} in ROSCIs (%)	DEV_{OLD_B} in ROSCIs (%)	Difference
−10 to 0%	No impact—slight abandonment	3.46	3.74	0.28
0 to 10%	No impact—minor overgrazing	3.22	4.41	1.19
−50 to −10.1%	Partial impact—moderate abandonment	28.86	15.44	−13.42
10.1 to 50%	Partial—moderate overgrazing	14.13	17.91	3.78
≤50%	Major impact (degraded)—abandonment	19.81	8.03	−11.77
>50%	Major impact (degraded)—overgrazing	30.52	50.46	19.93
No data	No data	0.01	0.01	0.00

Table 2.
Comparison of DEV_{OLD} percentages between the socioeconomic and biodiversity-focused scenarios in ROSCIs.

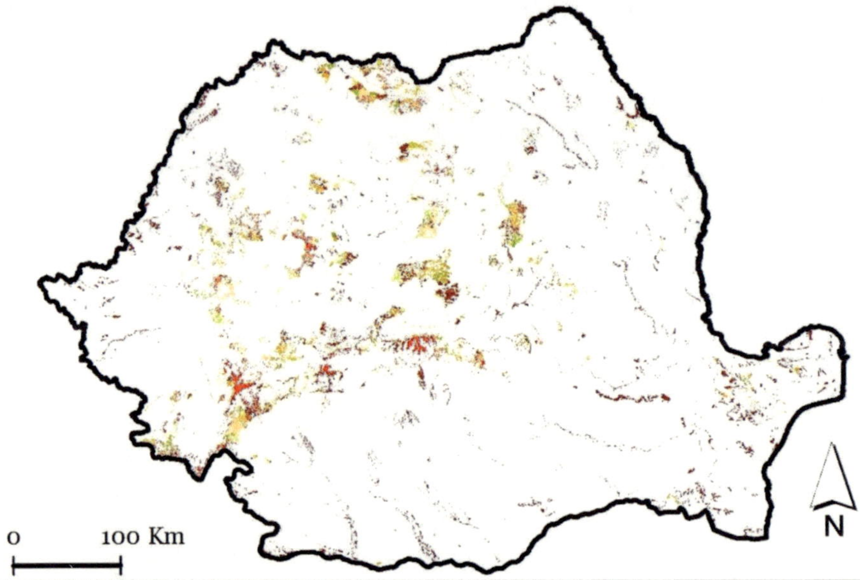

Deviation from Optimal Livestock Density for Biodiversity Conservation	Grassland Habitat Status in Natura 2000 sites (SCI) from Romania	Percent (%)	Area (km²)
-10% – 0%	No impact - Slight Abandonment	3.74	216.92
0% – 10%	No impact - Minor Overgrazing	4.41	256.24
-50% – -10.1%	Partial impact - Moderate Abandonment	15.44	896.47
10.1% – 50%	Partial - Moderate Overgrazing	17.91	1039.88
< -50%	Major impact (Degraded)- Abandonment	8.03	466.45
> 50%	Major impact (Degraded) - Overgrazing	50.46	2929.12
No Data	No Data	0.01	0.32

Figure 6.
The spatial distribution of impact and degradation within the N2000 ROSCIs caused by the deviations from grazing optimum for the biodiversity conservation scenario; the deviation classes, status, their percentage, and area at the national level.

However, most grasslands in the Eastern European socioeconomic region, similarly to other regions of Europe, within or outside the protected areas, were created and are maintained (along with their biodiversity) by an extensive form of management [4, 21, 59].

4. Conclusions

Since grassland habitats are very important for both socioeconomical and biodiversity reasons, their continuous degradation is a significant and urgent matter. Of the degradation factors, their use for forage/fodder and particularly the livestock density within the grasslands are highly relevant. This is also true for Romania, which hosts grassland habitats that are among the most extensive and diverse from the EU and where data regarding the impact of livestock density that deviates from the optimal value are not available. By combining environmental conservation data and agricultural statistics within a GIS, this paper assessed the grazing livestock density and the subsequent deviations from the optimal grassland carrying capacity within and outside the N2000 sites in order to highlight the areas with higher risk of impact and degradation and help monitor them. The extensive area analyzed, 33529.42 km^2, corresponds to all the permanent grasslands from Romania. The results indicate that more than half of this area is subject to a major impact and degraded, most of it from overgrazing. Only less than 10% of the permanent grassland area is not impacted by grazing livestock. Of the total national grassland area, 5815.75 km^2 (17.34%) of grasslands were determined to be situated within N2000 Sites of Community Importance, indicating the substantial presence within these protected areas of agricultural activities that are supported within the rural and regional European development policies. The major impact—degraded area—is slightly lower than at the national level by 50.34%, and within the N2000 grasslands abandonment is more important as an impact factor than at the national level. Given the high percentage of N2000 grassland habitats that are prone to major impact and degradation, the use of the lower, conservation-oriented optimal LD (of 0.45 LU/ha) is recommendable in their case. In this scenario, although the proportion of the strongly impacted-degraded N2000 is very similar (49.82%), the cause of degradation shifts toward a predominance of overgrazing, implying a need to reduce the livestock density in these areas. The simplest and most straightforward solution therein is to optimize the spatial distribution of the LD, particularly where neighboring TAUs experience opposite tendencies, abandonment vs overgrazing.

As a further approach, the spatial patterns of grazing intensity presented in the study allow to identify the areas where the intensity of agricultural practices is divergent from optimum and particularly where these and nature conservation efforts overlap. The detailed statistics obtained may serve as a basis for the design of optimized grazing and protection measures to prevent grassland degradation. This insight could further support policies aiming at a future conflict-free combination of agricultural production and nature conservation. The developed GIS toolbox can be used for environmental conflict anticipation and subsequent management of the grassland habitats so that both socioeconomic and conservation targets are achieved, being particularly useful in the case of protected areas. Although the analysis is focused on the Romanian grasslands, the model can be easily adapted to be used for similar situations abroad.

Acknowledgements

This study was supported by the following projects: SIPOCA22-Developing the administrative capacity of the Ministry of Environment to implement the biodiversity policies; PN-III-P4-ID-PCE-2016-0711 and Ctr. No. 22 PDI-PFE/2018 financed by the Ministry of Research and Innovation through UEFISCDI, and PN2019-2022/19270201-Ctr. No. 25N BIODIVERS 3-BIOSERV.

Conflict of interest

The authors declare no conflict of interest.

Notes/thanks/other declarations

Anamaria Roman, Tudor-Mihai Ursu, Irina Onțel, and József Pál Frink substantially contributed to the conception and design of this work. Teodor Marușca, Oliviu Grigore Pop, Sretco Milanovici, Alexandru Sin-Schneider, Carmen Gheorghe, Sorin Avram, and Sorina Fărcaș contributed to the data acquisition and interpretation. Anamaria Roman, Tudor-Mihai Ursu, and Irina Onțel performed the analysis and prepared the first draft of the manuscript. All authors participated in the final drafting of the work. All authors approve the final version of the work to be published.

Author details

Anamaria Roman[1], Tudor-Mihai Ursu[1*], Irina Onțel[2], Teodor Marușca[3], Oliviu Grigore Pop[4], Sretco Milanovici[5], Alexandru Sin-Schneider[6], Carmen Adriana Gheorghe[6], Sorin Avram[6,7], Sorina Fărcaș[1] and József Pál Frink[8]

1 Institute of Biological Research Cluj, National Institute of Research and Development for Biological Sciences, Cluj-Napoca, Romania

2 National Meteorological Administration, Bucharest, Romania

3 Grassland Research and Development Institute, Brașov, Romania

4 Foundation Conservation Carpathia, Brașov, Romania

5 National Museum of Banat, Natural Science Section, Timișoara, Romania

6 National Institute of Economic Research "Costin C. Kirițescu", Romanian Academy, Bucharest, Romania

7 Geography Department, University of Craiova, Craiova, Romania

8 National Institute for Research and Development in Forestry "Marin Drăcea", Cluj-Napoca, Romania

*Address all correspondence to: tudor.ursu@icbcluj.ro

References

[1] White R, Murray S, Rohweder M. Pilot Analysis of Global Ecosystems: Grassland Ecosystems. Washington, D.C.: World Resources Institute; 2000. p. 81

[2] Gibson DJ, editor. Grasses and Grassland Ecology. New York: Oxford University Press; 2009. p. 323

[3] Vitousek PM. Grassland ecology: Complexity of nutrient constraints. Nature Plants. 2015;**1**:1-2. DOI: 10.1038/nplants.2015.98

[4] Dengler J, Janišová M, Török P, Wellstein C. Biodiversity of palaearctic grasslands: A synthesis. Agriculture, Ecosystems and Environment. 2014;**182**:1-14. DOI: 10.1016/j.agee.2013.12.015

[5] Houlton BZ, Wang Y-P, Vitousek PM, Field CB. A unifying framework for dinitrogen fixation in the terrestrial biosphere. Nature. 2008;**454**:327-330. DOI: 10.1038/nature07028

[6] O'Mara FP. The role of grasslands in food security and climate change. Annals of Botany. 2012;**110**(6): 1263-1270. DOI: 10.1093/aob/mcs209

[7] Chang J, Viovy N, Vuichard N, Ciais P, Campioli M, Klumpp K, et al. Modeled changes in potential grassland productivity and in grass-fed ruminant livestock density in Europe over 1961-2010. PLoS One. 2015;**10**(e0127554): 1-30. DOI: 10.1371/journal.pone.0127554

[8] Hopkins A, Holz B. Grassland for agriculture and nature conservation: Production, quality and multi-functionality. Agronomy Research. 2006;**4**(1):3-20

[9] Werling BP, Dickson TL, Isaacs R, Gaines H, Gratton C, Gross KL, et al. Perennial grasslands enhance biodiversity and multiple ecosystem services in bioenergy landscapes. Proceedings of the National Academy of Sciences of the United States of America. 2014;**111**(4):1652-1657. DOI: 10.1073/pnas.1309492111

[10] Lehn F, Bahrs E. Land-use competition or compatibility between nature conservation and agriculture? The impact of protected areas on German standard farmland values. Sustainability. 2018;**10**(4):1-20. DOI: 10.3390/su10041198

[11] Squires VR, Feng H, Hua L. What future for the World's grasslands under global (not only climate) change? In: Squires VR, Dengler J, Feng H, Hua L, editors. Grasslands of the World: Diversity, Management and Conservation. London: CRC Press Taylor & Francis Group; 2018. pp. 390-406. DOI: 10.1201/9781498796262

[12] Wick AF, Geaumont BA, Sedivec KK, Hendrickson J. Grassland degradation. In: Shroder JF, Sivanpillai R, editors. Biological and Environmental Hazards, Risks, and Disasters. London: Elsevier; 2016. pp. 257-276. DOI: 10.1016/B978-0-12-394847-2.00016-4

[13] Gang C, Zhou W, Chen Y, Zhaoqi W, Sun Z, Li J, et al. Quantitative assessment of the contributions of climate change and human activities on global grassland degradation. Environment and Earth Science. 2014;**72**:4273-4282. DOI: /10.1007/s12665-014-3322-6

[14] Cai H, Yang X, Xu X. Human-induced grassland degradation/restoration in the central Tibetan plateau: The effects of ecological protection and restoration projects. Ecological Engineering. 2015;**83**:112-119. DOI: 10.1016/j.ecoleng.2015.06.031

[15] Andrade BO, Koch C, Boldrini II, Vélez-Martin E, Hasenack H,

Hermann JM, et al. Grassland degradation and restoration: A conceptual framework of stages and thresholds illustrated by southern Brazilian grasslands. Nature Conservation. 2015;**13**(2):95-104. DOI: 10.1016/j.ncon.2015.08.002

[16] Gartzia M, Pérez-Cabello F, Bueno GC, Alados C. Physiognomic and physiologic changes in mountain grasslands in response to environmental and anthropogenic factors. Applied Geography. 2016;**66**:1-11. DOI: 10.1016/j.apgeog.2015.11.007

[17] Wen L, Dong S, Li Y, Li X, Shi J, Wang Y, et al. Effect of degradation intensity on grassland ecosystem services in the Alpine Region of Qinghai-Tibetan plateau, China. PLoS One. 2013;**8**(3):1-7. DOI: 10.1371/journal.pone.0058432

[18] Isbell FI, Craven D, Connolly J, Bonin C, Bruelheide H, De Luca E, et al. Biodiversity increases the resistance of ecosystem productivity to climate extremes. Nature. 2015;**526**(7574): 574-577. DOI: 10.1038/nature15374

[19] Craine JM, Nippert JB, Elmore AJ, Skibbe AM, Hutchinson SL, Brunsell NA. Timing of climate variability and grassland productivity. Proceedings of the National Academy of Sciences of the United States of America. 2012;**109**(9):3401-3405. DOI: 10.1073/pnas.1118438109

[20] Zhang C, Wang X, Li J, Hua T. Roles of climate changes and human interventions in land degradation: A case study by net primary productivity analysis in China's Shiyanghe Basin. Environment and Earth Science. 2011;**64**:2183-2193. DOI: 10.1007/s12665-011-1046-4

[21] Török P, Janišová M, Kuzemko A, Rusina S, Stevanovic ZD. Grasslands, their threats and management in Eastern Europe. In: Squires VR, Dengler J, Feng H, Hua L, editors. Grasslands of the World: Diversity, Management and Conservation. London: CRC Press Taylor & Francis Group; 2018. pp. 64-88. DOI: 10.1201/9781498796262

[22] Papanastasis VP, Kyriakakis S, Kazakis G. Plant diversity in relation to overgrazing and burning in mountain Mediterranean ecosystems. Journal of Mediterranean Ecology. 2002;**3**(2):53-63

[23] Török P, Dengler J. Palaearctic grasslands in transition: Overarching patterns and future prospects. In: Squires VR, Dengler J, Feng H, Hua L, editors. Grasslands of the World Diversity, Management and Conservation. London: CRC Press Taylor & Francis Group; 2018. pp. 15-26. DOI: 10.1201/9781498796262

[24] Wiesmair M, Feilhauer H, Magiera A, Otte A, Waldhardt R. Estimating vegetation cover from high-resolution satellite data to assess grassland degradation in the Georgian Caucasus. Mountain Research and Development. 2016;**36**(1):56-65. DOI: 10.1659/MRD-JOURNAL-D-15-00064.1

[25] Başnou C, Pino J, Šmilauer P. Effect of grazing on grasslands in the Western Romanian Carpathians depends on the bedrock type. Preslia. 2009;**81**(2):91-104

[26] Chang J, Ciais P, Herrero M, Havlik P, Campioli M, Zhang X, et al. Combining livestock production information in a process-based vegetation model to reconstruct the history of grassland management. Biogeosciences. 2016;**13**:3757-3776. DOI: 10.5194/bg-13-3757-2016

[27] Kuemmerle T, Müller D, Griffiths P, Rusu M. Land use change in southern Romania after the collapse of socialism. Regional Environmental Change. 2009;**9**(1):1-12. DOI: 10.1007/s10113-008-0050-z

[28] Munteanu C, Kuemmerle T, Boltiziar M, Butsic V, Kaim D, Király G,

et al. Forest and agricultural land change in the Carpathian region—A meta-analysis of long-term patterns and drivers of change. Land Use Policy. 2014;**38**:685-697. DOI: 10.1016/j.landusepol.2014.01.012

[29] European Environmental Agency. Estimating the Environmentally Compatible Bioenergy Potential from Agriculture. EEA Technical Report. 2007. Available from: https://www.eea.europa.eu/publications/technical_report_2007_12

[30] Cooper T, Hart K, Baldock D. Provision of Public Goods through Agriculture in the European Union. EC Final Report. 2009. Available from: https://ec.europa.eu/agriculture/sites/agriculture/files/external-studies/2009/public-goods/report_en.pdf

[31] Neumann K, Elbersen BS, Verburg PH, Staritsky I, Perez-Soba M, de Vries W, et al. Modelling the spatial distribution of livestock in Europe. Landscape Ecology. 2009;**24**:1207-1222. DOI: 10.1007/s10980-009-9357-5

[32] Huyghe C, De Vliegher A, VVan Gils B, Peeters A, editors. Grasslands and Herbivore Production in Europe and Effects of Common Policies. Versailles: Quae; 2014. p. 323

[33] CLC2012. Corine Land Cover 2012 (CLC 2012); Romania. Country Fact Sheet. 2017; Romania. Available from: https://www.eea.europa.eu/themes/landuse/land-cover-country-fact-sheets/ro-romania-landcover-2012.pdf/view

[34] NSIR. Recensământul General Agricol 2010. 2011. Available from: http://www.insse.ro/cms/ro/content/recensamantul-general-agricol-2010

[35] Coldea G, Filipaș L, Fărcaș S, Stoica IA, Ursu TM, Pop A. The relationship between the structure of grasslands from the north-eastern slope of the Vlădeasa massif and socio-economic activities. Contributii Botanice. 2008;**43**:53-65

[36] Cristea V, Gafta D, Pedrotti F. Fitosociologie. Cluj-Napoca: Presa Universitară Clujeană; 2004. p. 394

[37] Vassilev K, Ruprecht E, Alexiu V, Becker T, Beldean M, Biță-Nicolae C, et al. The Romanian grassland database (RGD): Historical background, current status and future perspectives. Phytocoenologia. 2018;**48**(1):91-100. DOI: 10.1127/phyto/2017/0229

[38] Sutcliffe LME, Batáry P, Kormann U, Báldi A, Dicks LV, Herzon I, et al. Harnessing the biodiversity value of central and eastern European farmland. Diversity. 2015;**21**:722-730. DOI: 10.1111/ddi.12288

[39] Doniță N, Popescu A, Paucă-Comănescu M, Mihăilescu S, Biriș AI, editors. Habitatele din România. București: Editura Tehnică Silvică; 2005. p. 496

[40] Gafta D, Mountford JO, editors. Manualul de Interpretare a Habitatelor Natura 2000 din România. Cluj-Napoca: Risoprint; 2008. p. 104

[41] Coldea G, Oprea A, Sârbu I, Culiță S, Ștefan N. Les Association Végétales de Roumanie (Tome 2. Les associations anthropogènes). Cluj-Napoca: Presa Universitară Clujeană; 2012. p. 482

[42] Sârbu A, Coldea G, Negrean G, Cristea V, Hanganu J, Veen P. Grasslands of Romania. National Grasslands Inventory 2000-2003; University of Bucharest & KNNV. Bucharest: Ediura Alo; 2004. p. 58

[43] ESRI. ArcGIS Version 10.5. 2017

[44] LPISweb-PIAA. Romanian Land Parcel Identification System. Romania: Paying and Intervention Agency For Agriculture; 2017. Available from: http://www.apia.org.ro/files/

pages_files/17-03-23-01-14-22GHID_utilizare_IPA_ONLINE_2017.pdf

[45] Madr O. Ordinul nr. 544/2013 Privind Metodologia de Calcul al Încărcăturii Optime de Animale pe Hectar de Pajişte. Monitorul Oficial2013. pp. 3-6. Available from: https://lege5.ro/Gratuit/gm3tcnzrha/ordinul-nr-544-2013-privind-metodologia-de-calcul-al-incarcaturii-optime-de-animale-pe-hectar-de-pajiste

[46] Maruşca T, Mocanu V, Cardaşol V, Hermenean I, Blaj VA, Oprea G, et al. Ghid de Producere Ecologică a Furajelor de Pajişti Montane. Braşov: Editura Universităţii Transilvania din Braşov; 2010. p. 176

[47] Maruşca T, Mocanu V, Haş EC, Tod MA, Andreoiu AC, Dragoş MM, et al. Ghid de Întocmire a Amenajamentelor Pastorale. Braşov: Editura Copalavoro Braşov; 2014. p. 248

[48] Bărbos MI, Frink JP, Maruşca T, Milanovici S, Onţel I, Sin A. Cartarea ecosistemelor de pajişti naturale şi seminaturale degradate din România. In: Avram S, Croitoru A, Gheorghe C, Manta N, editors. Cartarea Ecosistemelor Naturale şi Seminaturale Degradate. Bucureşti: Editura Academiei Române; 2018. p. 232

[49] Maruşca T, Bărbos MI, Blaj VA, Cardaşol V, Dragomir N, Mocanu V, et al. Tratat de Reconstrucţie Ecologică a Habitatelor de Pajişti şi Terenuri Degradate Montane. Braşov: Editura Universităţii "Transilvania"; 2010. p. 359

[50] Barcella M, Filipponi F, Assini S. A simple model to support grazing management by direct field observation. Agriculture, Ecosystems and Environment. 2016;**234**:107-117. DOI: 10.1016/j.agee.2016.04.027

[51] Török P, Hölzel N, van Diggelen R, Tischew S. Grazing in European open landscapes: How to reconcile sustainable land management and biodiversity conservation? Agriculture, Ecosystems and Environment. 2016;**234**:1-4

[52] Török P, Valkó O, Deák B, Kelemen A, Tóth E, Tóthmérész B. Managing for species composition or diversity? Pastoral and free grazing systems in alkali steppes. Agriculture, Ecosystems and Environment. 2016;**234**(2015):23-30. DOI: 10.1016/j.agee.2016.01.010

[53] Török P, Penksza K, Tóth E, Kelemen A, Sonkoly J, Tóthmérész B. Vegetation type and grazing intensity jointly shape grazing effects on grassland biodiversity. Ecology and Evolution. 2018;**8**(20):10326-10335. DOI: 10.1002/ece3.4508

[54] Tóth E, Deák B, Valkó O, Kelemen A, Miglécz T, Tóthmérész B, et al. Livestock type is more crucial than grazing intensity: Traditional cattle and sheep grazing in short-grass steppes. Land Degradation & Development. 2018;**29**(2):231-239. DOI: 10.1002/ldr.2514

[55] Temme AJAM, Verburg PH. Mapping and modelling of changes in agricultural intensity in Europe. Agriculture, Ecosystems and Environment. 2011;**140**(1-2):46-56. DOI: 10.1016/j.agee.2010.11.010

[56] Schulp CJE, Levers C, Kuemmerle T, Tieskens KF, Verburg PH. Mapping and modelling past and future land use change in Europe's cultural landscapes. Land Use Policy. 2019;**80**:332-344. DOI: 10.1016/j.landusepol.2018.04.030

[57] Milchunas DG, Lauenroth WK. Quantitative effects of grazing on vegetation and soils over a global range of environments. Ecological Monographs. 1993;**63**(4):327-366. DOI: 10.2307/2937150

[58] Yang X. Assessing responses of grasslands to grazing management using

remote sensing approaches [thesis]. Saskatoon, Saskatchewan: University of Saskatchewan; 2012

[59] Loss J, Turtureanu PD, von Wehrden H, Hanspach J, Dorresteijn I, Frink JP, et al. Plant diversity in a changing agricultural landscape mosaic in southern Transylvania (Romania). Agriculture, Ecosystems and Environment. 2014;**199**:350-357. DOI: 10.1016/j.agee.2014.10.013

Chapter 4

Mangrove Habitat Loss and the Need for the Establishment of Conservation and Protected Areas in the Niger Delta, Nigeria

Aroloye O. Numbere

Abstract

Niger Delta mangroves are the largest in Africa, but uncontrolled anthropogenic activities had reduced their population size. The reduction from large to small mangrove stand has some ecological implications on species populations. For instance, stochastic events such as flooding, landslides, sea level rise, high temperature, and humidity affect small populations. Human-mediated actions of random deforestation for firewood production, canalization, and de-silting of waterways, lead to the complete elimination of mangrove stands in specific locations. The cumulative effect of these actions can result in local extinction and loss of genetic variation of mangroves. Destruction of mangroves over the years is detrimental to other species that inhabit the mangroves in the Niger Delta (e.g., fishes, crabs, etc.). This situation can be reversed or stopped if effective protective measures are adopted. Strict protective measures can be done in areas that are highly impacted i.e., regions where oil and gas exploration or massive deforestation activities had occurred. Limited protection can be done in areas with low impact, and is known as a win-win conservation where the peoples welfare is considered. Here, indigenous people are employed to help in the protection of the forest and in return are allowed to exploit its resources.

Keywords: mangrove, Niger Delta, fire wood, dredging, hydrocarbon pollution, urbanization, deforestation, maritime transport

1. Introduction

Mangroves are shrubs or small trees that inhabit the interface between the land and the sea. They are habitat specialists that grow along coastal regions. They grow in swampy soil that is continuously wet from year to year because of the action of tides and heavy rainfall. Mangroves are found in tropical areas with high solar radiation and precipitation [1]. The mangroves of the Niger Delta are the third largest in the world and the largest in Africa [2]. It is a biodiversity hot spot, and is found in the tropics close to the equator where rainfall persists all year round. The climate is hot and facilitates the proliferation of living organisms from microbes to vertebrates through series of speciation and evolutionary events. Biodiversity hot spots are important zones because they contain 25% of all terrestrial species. Hot spots are small areas on the map (i.e., about 2%), but wield so much ecological influence on the surface of the earth. They contain disproportionately large numbers of species on earth. Tropical forest has the largest number of species in the world. It has a lot of taxonomic diversity, which includes plankton, macrophytes, arthropods, fungi, protists, mollusks, crabs, vertebrates including birds and mammals. Mangrove forest is the most dominant amongst the plants found in the coastal region of the Niger Delta area. The reasons for high species diversity in mangrove forests of the Niger Delta are because of high reproduction and productivity [3]. The tropics are areas that have high speciation and less species extinction. Secondly, the stability hypothesis postulates that the tropics have been more stable than present in the last 1000 years and will remain so for a long period of time. It is stable with respect to the abiotic conditions that prevail in the area. Similarly, the equator has been stable with respect to global warming and cooling with lower extinction rate recorded over the years. Furthermore, intense solar radiation had resulted in an increase in overall productivity resulting in high biodiversity turnover. The tilt of the earth also affects how the photon of sun hits the earth. The sun position right overhead the equator serves as a major source of energy. Plants capture solar energy, which they convert to food for other organisms in the ecosystem. The captured energy is converted into biomass, which is transferred to the production of more resources. This is exemplified in the high productivity of mangrove forests recorded in the Niger Delta [3]. Increased litter fall facilitates high decomposition [1] and makes food available for different species of organisms in the food chain. The tropical environment is highly specialized. High specialization and speciation lead to high biodiversity. But the problem of mangroves globally is the gradual loss of their habitat due to anthropogenic activities [4]. This study showed that 35% area of global mangrove forest had been lost as a result of some human activities such as shrimp culture, forest use, fish culture, diversion of freshwater, land reclamation, herbicides, agriculture, salt ponds and coastal developments. In Africa coastal development is a major factor of habitat loss [4]. Although, other studies had revealed that Nigeria has one of the least carbon CO_2 emissions from soil as a result of mangrove losses [5]. Similarly, previous studies have shown that habitat conversion far exceeds habitat protection by a ratio of 8:1 globally [6].

1.1 Mangrove forest resource in Nigeria

A forest is an ecosystem, which is dominated by trees. Forest produces timber and non-timber products. Non-timber forest includes anything other than timber. Forest products are all renewable and can be sustainably managed. More often than not mangrove forest is looked at from the economic resource angle. This is because they are the sources of revenues such as wood, firewood, latex, dyes, thatches, bamboo, reptiles, insects, roots, shoots, stems, flowers, honey resins, gum, silk, fabrics, rope, animal oil, cosmetics, water from streams and lakes. The value of these resources is often difficult to quantify. We also have nature reserves and recreational activities.

Nigeria has a wide and diverse range of habitats from arid zones in the north to the swampy wetlands in the south. Many forest types (tropical rainforest, Sahel savanna, Guinea savanna, Sudan savanna, Montane savanna, etc.) are associated with these zones and have an array of plant and animals species. Some of these biodiversity include:

1. **Plants**: There are over 4,600 plant species in Nigeria. Out of these figures 205 plants species are endemic. But they are under the threat of extinction through due to overexploitation.

2. **Mammals**: There are about 274 species of mammals, which is the highest number in Africa. Two endemic species are the white coated guenon and the red tailed monkey. They are also endangered.

3. **Birds**: Nigeria has about 839 species of birds with two as endemic. They are Anambra waxbill and Ibadan malimbe. All birds of prey are considered threatened in Nigeria

4. **Reptiles**: They are about 56 species of forest snakes in Nigeria. We also have 56 savanna snake species. Crocodiles, pythons, monitor lizards and turtles, which are under threat.

5. **Amphibians**: There are about 19 recovered species of amphibians in Nigeria, 5 of which are endemic to Nigeria and Cameroun, and 3 are under threat of extinction from habitat destruction.

6. **Invertebrates**: The giant swallow tail and generally other lepidopteran species are under threat in Nigeria due to pollution from oil and gas exploration. Only about 1.7% of Nigeria land is included within protected area as compared with the large landmass. Even the protected areas are being degraded by poaching, illegal logging and infrastructural development.

2. Habitat loss and conversion

There is a gradual but steady loss of mangrove forests in the Niger Delta due to uncontrolled deforestation for the purpose of sand dredging and canalization (**Figure 1**). Mangrove forest is also cut to recover stems, which are used in the production of firewood and wood for the construction of houses. Similarly, numerous oil and gas exploratory activities all over the Niger Delta area open up the forests to further exploitation of resources and invasion by foreign species such as nypa palms (*Nypa fruticans*) [7]. Habitat loss is amongst the three important factors that are responsible for recent species extinction. The other two are overexploitation and the introduction of exotic species [8]. The implication of the loss of mangrove habitat is the loss of ecosystem services it renders to the society [9].

2.1 Habitat loss

Habitat loss leads to extinction because species are adapted to a particular habitat type. For instance, mangroves are habitat specialist that survives in swampy wetland soil, therefore, when they are removed from their native environment and taken to other environment they do not survive. There is also a relationship between the numbers of habitats and the species present. This is because the larger the habitat size the higher the number of species, and the smaller the habitat size the fewer the number of species present. There are many ways humans cause the loss or conversion of mangrove forests, these include tree cutting for production of firewood, agriculture (e.g., fish farming or rice paddies), construction of houses and hydrocarbon pollution from oil and gas exploration activities. Currently, there are numerous

Figure 1.
Consequences of biodiversity loss on species population: (A–C) shows deforestation for fire wood production; (D–F) show canalization of river bottom to create way for sea craft; and (G–I) show the displacement and migration of birds (heron) to urban areas for breeding due to the loss of their natural habitat in the Niger Delta, Nigeria.

locations in the Niger Delta where sand dredging and mining activities are taking place. Usually before the sand dredging the entire mangrove forests in such vicinity are cut down to pave way for the entry of heavy duty machinery (e.g., **Figure 1E, F**).

2.2 Habitat conversion

It is a situation whereby mangrove forests are converted into sand fill or other land use systems that are inimical to the existence and proliferation of coastal species. For instance, in sand filled land only weed grow as the main vegetation cover. This leads to a drastic change in the total number of species. In this context, loss of mangrove forests lead to the loss of forest canopy that houses other species such as birds, tree crabs, monkeys, etc. Increased human activities such as industries, living quarters, and marine transportation result in the increase in waste disposal in aquatic environment. The disposal of liquid and solid wastes into the river contaminates adjoining mangrove forests. Accumulation of organic products promotes the proliferation of algae, which accelerate the process of eutrophication. Fertilization of the aquatic system leads to the increase in biological oxygen demand (BOD) and decrease in dissolved oxygen (DO) in the water. Accumulation of organic waste in mangrove forest leads to anaerobic condition which causes slow death of mangroves. The implication of these activities is that once the habitat type is converted protection becomes extremely difficult [6].

2.3 Mangrove habitat fragmentation

Mangrove fragmentation is the process whereby large contiguous areas of mangrove forests are divided into small habitats as a result of urban development [10].

Figure 2.
*Isolated and fragmented mangrove stands surrounded by invasive nypa palm species (*Nypa fruticans*) and an overlooking urban area at Eagle Island, Niger Delta Nigeria. This mangrove stands are subject to impending stochastic events such as flooding, tidal surge, sediment erosion at the root of the plants which may easily wipe it out in few years time.*

When this happens, fragmented forests become physically isolated from each other, which has a negative effect on the population dynamics of organisms in the forest (**Figure 2**). Fragmentation of mangrove forests has an effect on the connectivity of the forest, which affects the movement of organisms within the forest. The island biogeography theory vividly throws more light on this situation when [11] in his seminal work showed that big island has more species than small islands. In the same way, big mangrove forest will have more biodiversity and less extinction than small mangrove forests. Furthermore, island biogeography theory shows that island close to main land has more species than island far away from mainland. In contrast, the closer the mangrove forest is to urban area the lower the number of species due to human disturbances. Mangroves are only adapted to coastal areas, which prevent them from migrating upland [12]. However, their seeds can migrate to foreign lands through tidal pressure far away from their place of origin. Some individual mangrove seedlings can migrate to oceanic habitat patches. The distance between the patches will affect the rate at which the species get there. This is because human activities that lead to land fragmentation such as urban development and sand filling can lead to the compartmentalization of the mangrove forests. The mangrove in the Niger Delta area has become a sink population, and thus there is a negative population growth due to anthropogenic activities.

2.3.1 Impact of edge effect and fragmentation on mangrove forest

Edge is a location of an abrupt transition between two habitat types. When a large area is fragmented, it increases the amount of edges on a habitat and decreases core habitat, thus exposing the edges to more impacts by humans who gain access to the forest to cause more plundering of its resources. An example is the entry of

Climate/human parameters	Core mangrove habitat	Edge mangrove habitat
Temperature	Lower	Higher
Humidity	Higher	Lower
Light level	Lower	Higher
Wind	Lower	Higher
Human impact	Lower	Higher

Table 1.
The level of environmental impact on fragmented mangrove forests in the Niger Delta, Nigeria.

invasive nypa palm species into mangrove forest due to the direct introduction by humans [7]. Fragmentation of large mangrove forest into small fragments lead to the formation of little mangrove islands that are prone to forces of denudation such as tidal pressure, erosion and flash floods which erodes it banks leading to a gradual loss of the entire mangrove stands (**Figure 2**).

The difference between core and edge mangrove forests in relation to habitat loss is because of the impact of some climatic parameters (**Table 1**).

A major cause of mangrove deforestation in the Niger Delta, Nigeria is its use in the production of firewood, which reduces mangrove forest sizes. The fragmentation of mangrove forest can lead to the formation of numerous edges that subject the forest to further climatic and anthropogenic damages.

2.4 Rarity of species

Rarity of species can be described in three ways [13, 14]:

1. Geographic range: it means how much of landscape is covered by the mangrove forest. It can be wide (i.e., found over a large area), or narrow (i.e., it will exhibit a rare form of the habitat size).

2. Habitat specificity: this involves whether the forest is narrow or restricted to coastal area or broad and found in different habitat type such as riverine and upland locations.

3. Population size: this is important because large population of a given species is better than medium or small populations of species. This is because extinction risk increases as population size declines.

2.5 Danger of small populations of mangrove forests

There are basically six problems associated with small populations of any kind of species [15], they are:

1. **Environmental stochasticity**: there is no constant environment, because of the action of fluctuation. Environmental stochasticity is an unpredictable event such as changes in weather (i.e., climate variability), invasive species, parasites and diseases, soil nutrients, etc. that has to do with environmental variation, and causes variation in survivorship of individuals. If population is large, it is not threatened with the possibility of extinction. But if the population is small it has high probability of extinction. The stochasticity, which

affects survival affects small population size more, which eventually leads to zero. Moderate population size will also go to zero with time. This can occur during major disturbances such as flood, mudslide, deforestation, canalization, sand filling and sand dredging activities. Deforested mangrove will also suffer from reduced population, which will further lead to habitat loss.

2. **Demographic stochasticity**: this is the chance of fluctuation or randomness that is common in small populations. Small populations of mangroves are at a risk of low seedling turnover, which leads to low growth rate. It is easier for small mangrove stands to be wiped out due to environmental and anthropogenic factors than large mangrove stand.

3. **Inbreeding depression**: this can occur in mangroves especially when population size drops resulting in reduced fitness as a result of cross pollination between similar species. This is because individuals in small inbred populations may have small germination rate when compared to large inbred populations. This situation is observed in two mangrove forest communities situated in Buguma and Eagle Island in the Niger Delta. There is a rejuvenated and increased germination rate in red and black mangroves in Buguma as compared to Eagle Island. This is because the mangrove forest in Buguma is isolated from human disturbances resulting in high canopy cover whereas mangroves in Eagle Island are scanty with few populations because of poor growth rate as a result of human intrusion into the mangrove forest. Low population size can also lead to loss of genetic variation.

4. **Genetic variation:** The theory of natural selection states that the rate of evolutionary change in a population is proportional to the amount of genetic variation in the population. Based on this hypothesis all populations can respond quicker to any change in the environment where they have higher genetic variation than lower genetic variation. Populations with high genetic variation can easily adapt to any change. Although, this situation operates at a long term scale. The mangroves are resilient species and take between 20 to 50 years to attain maturity, therefore, any genetic change in mangroves based on environmental perturbations such as pollution from oil spillage will take along time to manifest in the population.

According to [16], there is a relationship between population size and a probability of genetic variation that would be lost from one generation to the next. This is exemplified in Eq. (1):

$$H = 1 - \frac{1}{2N_e} \tag{1}$$

where H is the proportion of heterozygosity remaining in population from one generation to the next and N_e is the effective population size.

Effective population size is the number of adults breeding that contributes to equal genetic material in the population. This implies that small populations tend to lose heterozygosity over time, which is an argument for maintaining large populations and thus larger reserves wherever possible to conserve these species. Years of population abundance can be interrupted by stochastic events leading to population bottlenecks, a situation, which occurs when a population experiences a severe temporary reduction in size. Thus population bottleneck reduces effective population size, thus, when population size is small genetic variation declines. Small population size creates extinction vortex, where abundant population continuously gets smaller and finally goes extinct.

3. Estimation of population viability

To understand the population dynamics of mangrove forest in the face of anthropogenic action, the minimum viable population size (MVP) needs to be estimated. It is defined as the smallest isolated population size that has 99% survival probability for 100 years period [17]. It is a crucial tool of conservation for the prevention of extinction of species. For perennial plant such as mangroves, it should have a minimum population size of 500–5000 individuals in a given location from year to year. This means any isolated forest below 500 tree stands is not a viable population and has the risk of going into extinction within a short time would be high. Two methods can be employed in determining the MVP. They are:

1. Study of viability of population at different sizes: this method is like a trial and error where different isolated populations are observed for some time to see if there will be survival. If the tree stands survive they are viable, but if they do not survive they are not viable.

2. Population viability analysis (PVA): this method is a mathematical model that can be applied to studying mangrove population. Here assumptions are made to simplify factors that cause population decline. This method can be used for any species to determine the viability of population. Two types of PVA approaches include: (i) count-based PVA: this involves the collection of information from a number of individuals in a population for a period of time. The steps include: (a) counting the population size for some time (b) calculating population growth rate each year, and (c) constructing a model that predicts future population size based on the growth obtained. (ii) Demographic PVA: it uses demographic information explicitly in the PVA. It incorporates life history data into the PVA.

4. Establishing protected areas around mangrove forest in the Niger Delta

The mangrove forest in the Niger Delta needs urgent protection due to the adverse impact of human activities, which had resulted in the decimation of their populations in many locations. Unabated destruction of the mangrove forest will result in local extinction of these species. Therefore, to reduce human impact on mangroves, protected areas need to be urgently established to restrict human entry into forest. This can be achieved in three ways:

1. By law (public lands) for state or federal level

2. By purchase of private lands

3. By conservation easement, which would restricts developmental rights within mangrove forest. Deeds of property to land owners within and around mangrove forests should be revoked or restricted to prevent the destruction of mangrove forest for purpose of building or establishing of any kind of developmental project (e.g., resort). Furthermore, persons who own land within this area can be compensated and asked to relinquish ownership for the sake of conserving the forests.

Restricted areas can be declared by both state and federal government as national park to prevent the exploitation of its resources by land speculators. Inspiration can be drawn from the first national park i.e. the Yellowstone National Park, that was established in the United States, in 1872 by President Grant [18]. It measures 300 square miles and had geological features, which is of national interest. As for mangrove forests in the Niger Delta there are no major natural features therein that makes them to attract the attention of government for protection. Nevertheless, the interest in the mangrove forest in the Niger Delta lies in the extraction of their resources such as oil and gas exploitation and exploration, tree felling for firewood production and sand dredging activity. Although the mangroves have no geological significance it has ecological significance because of the numerous ecosystem services they render to the environment (e.g., coastal protection from tidal flushing, biodiversity hotspot).

The Niger Delta mangrove forest is a global biodiversity hotspot. Global biodiversity hotspots are often determined based on species richness and endemism, which are all found in the mangroves of the Niger Delta. These areas include: tropical rainforest, coral reef, alpine forest and Mediterranean areas. Four notable global biodiversity hotspots based on the assessment of the author are:

1. Niger Delta mangrove forest zones: these are areas in Nigeria that have the highest concentration of aquatic organisms in Africa (**Figure 3**). The problem is that the area has not been formally recognized as a global biodiversity hotspot by international agencies such as the the International Union for the Conservation of Nature (IUCN). The species range from plankton (phytoplankton and zooplankton), aquatic invertebrates (bivalves, crabs, mussel, periwinkles, hermit crabs, etc.), land insects (beetles, butterflies, mosquitoes, ants and termites), and vertebrates (monkeys, manatee, hippopotamus etc). It has five major species of mangroves amongst others. These include: red mangroves (*Rhizophora* spp.), black mangroves (*Avicennia* spp.), white mangroves (*Laguncularia germinans* spp.), buttonwood (*Conocarpus* spp.) and mangrove fern (Acrostichum aureum).

2. Cape floristic province: it has the greatest non-tropical concentration of higher plants species in the world. Five of South Africa's 12 endemic plant families are found in the Fynbos. One hundred and sixty species are endemic.

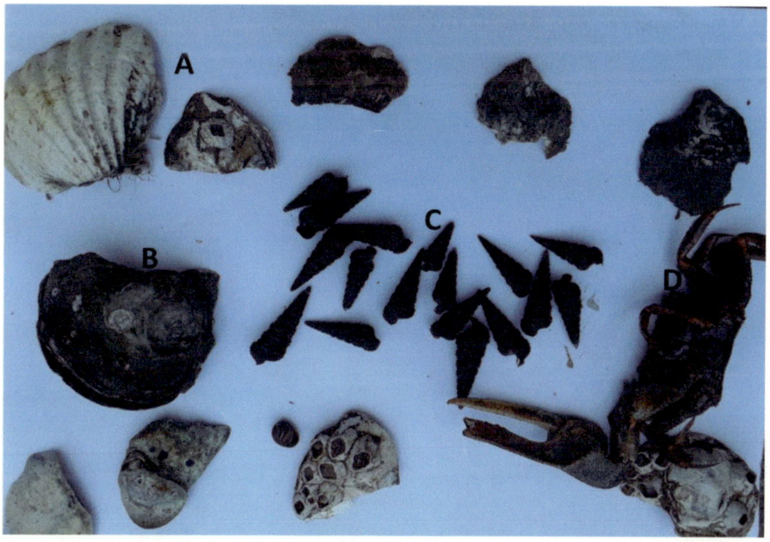

Figure 3.
Some species found within the Niger Delta mangrove forest: (A) Senilia senilis, *Anadara; (B)* Crassostrea gasar, *oyster; (C)* Tympanotonus fuscatus, *periwinkle; and (D)* Uca tangeri, *male crab.*

3. California Floristic Province: has the largest avian breeding ground in the US. It has large number of endemic species with many threatened. It is the source of all agricultural production in the US.

4. Mountains of Southwestern China: it is the most endemic rich temperate flora in the world. Golden monkey, giant panda, red panda and snub nose monkeys are found in this area.

Biodiversity hotspots in Africa are more often grouped together without the recognition of the rich biodiversity across different locations. For instance, the entire forest biodiversity in West Africa was grouped as "West African forests" by [19]. Whereas, each country in this region possess rich supply of biodiversity. Niger Delta area in Nigeria has the third largest mangrove forest in the world, and the largest in Africa, i.e., 1 million hectare out of 3.2 hectare in the whole of Africa.

Based on the importance of the biodiversity hotspots to the environment, it is pertinent to protect them for future generation. This is because only 6% of the earth's surface is protected. Half of these are scientific reserves and national parks (1.3%). They vary greatly from country to country. There are already six major parks in Nigeria [7]. But there is a need for mangrove forest parks in critical areas of the region aimed at forestalling the degradation of the forest. Two kinds of protection to be considered (**Figure 4**) are

1. **Strict protection**: in this type of protection, no resource extraction is allowed. Here trees will not be felled for firewood production and aquatic organisms will not be harvested for commercial or subsistence purposes. The sole aim of this kind of protection is biodiversity conservation. However, there might be some allowances for scientific and educational uses to increase knowledge in mangrove research.

2. **Limited protection**: Here some resource extraction can be allowed such as hunting, fishing, logging, use of stem for firewood production, tourism and recreation, etc. In this type of protection there is much more human impact, but the primary goal is the management of the natural resources for multiple goals.

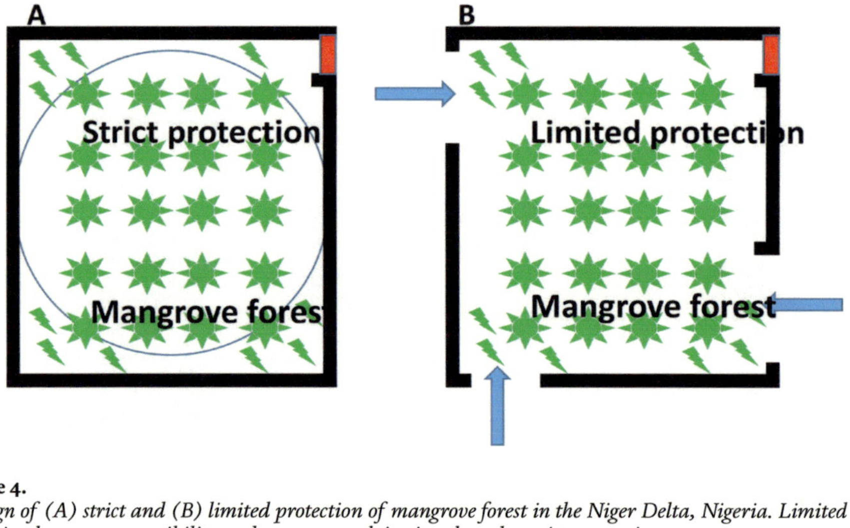

Figure 4.
A design of (A) strict and (B) limited protection of mangrove forest in the Niger Delta, Nigeria. Limited protection has more accessibility and resource exploitation than the strict protection.

In the past, conservation of biodiversity hotspot was mainly focused on species richness, but now it considers ecosystem integrity, water quality, climate impacts, unique adaptations, ecosystem services, intact fauna and flora, specialized/unique habitat and ecological processes.

In other to establish priority for protecting mangrove zones, the following need to be considered:

3. The goal of a particular protected area:

   a. what to preserve in terms of species, community and ecosystem? Here the mangrove species (red, black and white) are to be protected along with other flora and fauna communities and the entire ecosystem in identified locations. In this type of protection, firstly, species approach and historical records can be used to determine endangered species, taxonomic uniqueness and endemic species. Secondly, the community and ecosystem approach (modern tactics) can be used, which focuses on protecting the ecosystem to preserve the communities and species within. In this approach, the species and ecosystem are not mutually exclusive. This is because if the ecosystem is healthy and intact, the species will equally be healthy and intact.

   b. where to preserve? (a) Gap analysis can be used: It is the use of various remote sensing data to build overlaid sets of maps of various parameters (e.g., vegetation, soils, protected areas, species distribution) to identify spatial gaps in species protection and management programs (**Figure 5**). This can be performed at local, regional and global levels. It is done for both threatened and common species. It is typically done using GIS via map overlays. However, the use of gap analysis is becoming old-fashioned because currently small drones called unmanned aerial vehicles (UAV) are now deployed to study forest including protected and unprotected areas (**Figure 6**). It can also be used to study the impact of invasion on mangrove forest [20].

Figure 5.
A three-dimensional reconstruction of tree canopy and gaps within some mangrove forests captured with a DJI drone at Eagle Island, Niger Delta, Nigeria.

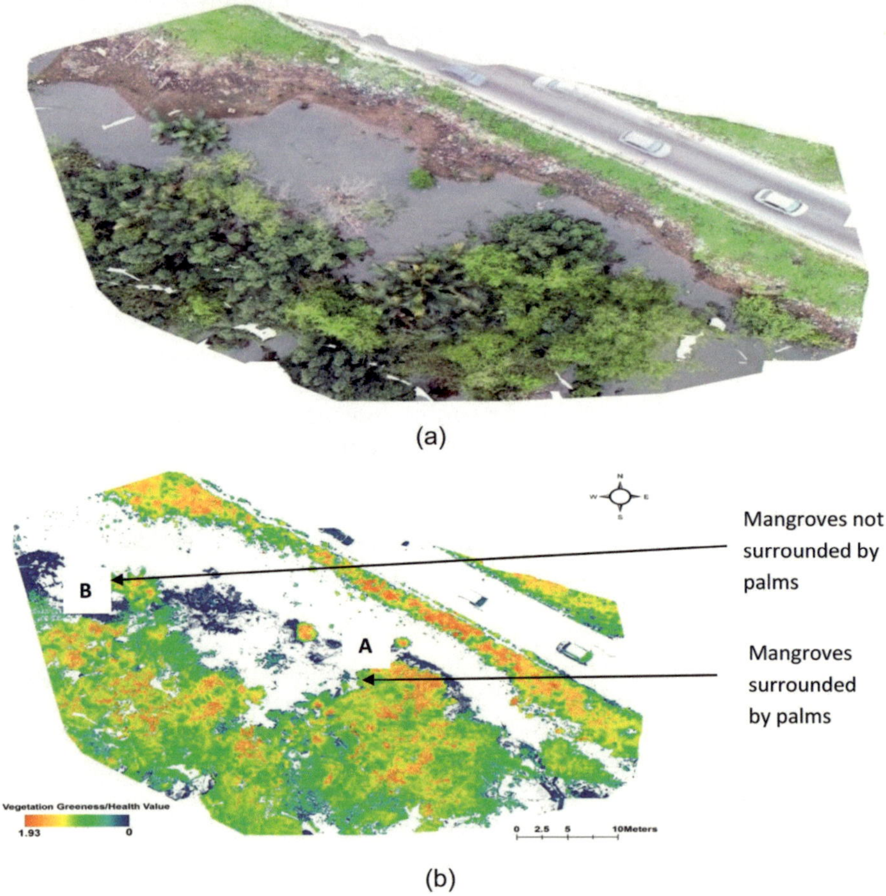

Figure 6.
(a) RGB imagery of mosaicked mangrove forest at Eagle Island, Niger Delta Nigeria (Source: [20]). (b) Mosaicked image of mangrove forest that has been processed with visible atmospherically resistance index (VARI) to show areas of stress caused by Nypa palms and depicted by red and yellow color (Source: [20]).

The protection of mangrove forest should follow the principles of reserve design such as:

1. Protection of entire habitats because the more the protected habitat the better it is.

2. Avoidance of reserve fragmentation by anthropogenic activities such as construction of high ways through mangrove forest.

3. Clumped reserve is far better than linear for easy migration of species within the mangrove forests

4. Circular reserve is ideal to minimize edge effects.

In terms of size, the bigger the terrain the better for the proliferation of species. According to the species-area relationship, bigger reserve has more resource, greater population with greater biodiversity leading to lower probability of extinction [21]. This is because small populations are prone to more extinction of species than large populations. Similarly, rather than have one large reserve it is good to have several small reserves (SLOSS problem), which will protect more species. This is because isolation can stop the spread of diseases.

5. Conclusion

Habitat loss and conversion are two major problems that can lead to the extinction of mangrove forest in the Niger Delta if not checked. This is because one or two stands of mangrove forest, which is made up of at least 5–10 trees are lost daily from this region as a result of deforestation for firewood (i.e., logging), sand dredging, urban development (e.g., roads, building of houses), etc. As the mangroves are brought down, their positions are quickly taken over by human structures such as roads, houses, industrial complexes and crude oil platforms. Areas that have not undergone infrastructural development, but have been disturbed by human actions contain scanty forests that become vulnerable to environmental pressures from invasive species, which had already completely taken over 60% of mangrove forest in the Niger Delta. Urgently, it is important to embark on deliberate protective measures, which can prevent the exploitation and plundering of the remaining mangroves resources in the zone.

Author details

Aroloye O. Numbere
Department of Animal and Environmental Biology, University of Port Harcourt, Choba, Nigeria

*Address all correspondence to: aroloyen@yahoo.com

References

[1] Numbere AO, Camilo GR. Mangrove leaf litter decomposition under mangrove forest stands with different levels of pollution in the Niger River Delta, Nigeria. African Journal of Ecology. 2016;**55**:162-167

[2] World Bank. Defining an Environmental Management Strategy for the Niger Delta. Report prepared by Jsdip Smig, David Moffat and Olof Linden. Washington DC: Industry and Energy Operations Division, West African Department, World Bank; 1995

[3] Numbere AO, Camilo GR. Structural characteristics, above-ground biomass and productivity of mangrove forest situated in areas with different levels of pollution in the Niger Delta, Nigeria. African Journal of Ecology. 2018;**54**(4):917-927

[4] Valiela I, Bowen JL, York JK. Mangrove forests: One of the World's threatened major tropical environments: At least 35% of the area of mangrove forests has been lost in the past two decades, losses that exceed those for tropical rain forests and coral reefs, two other well-known threatened environments. Bioscience. 2001;**51**:807-815

[5] Atwood TB, Connolly RM, Almahasheer H, Carnell PE, Duarte CM, Lewis CJE, et al. Global patterns in mangrove soil carbon stocks and losses. Nature Climate Change. 2017;**7**:523

[6] Hoekstra JM, Boucher TM, Ricketts TH, Roberts C. Confronting a biome crisis: Global disparities of habitat loss and protection. Ecology Letters. 2005;**8**:23-29

[7] Numbere AO. The impact of oil and gas exploration: Invasive nypa palm species and urbanization on mangroves in the Niger River Delta, Nigeria. In: Makowski C, Finkl C, editors. Threats to Mangrove Forests. Coastal Research Library, Vol. 25. Cham: Springer; 2018

[8] Aber J, Neilson RP, McNulty S, Lenihan JM, Bachelet D, Drapek RJ. Forest processes and global environmental change: Predicting the effects of individual and multiple stressors: We review the effects of several rapidly changing environmental drivers on ecosystem function, discuss interactions among them, and summarize predicted changes in productivity, carbon storage, and water balance. Bioscience. 2001;**51**:735-751

[9] Dobson A, Lodge D, Alder J, Cumming GS, Keymer J, McGlade J, et al. Habitat loss, trophic collapse, and the decline of ecosystem services. Ecology. 2006;**87**:1915-1924

[10] Wang P, Numbere AO, Camilo GR. Long term changes in mangrove landscape of the Niger River Delta, Nigeria. American Journal of Environmental Sciences. 2016;**12**:248-259

[11] Zimmerman BL, Richard OB. Relevance of the equilibrium theory of island biogeography and species-area relations to conservation with a case from Amazonia. Journal of Biogeography. 1986;**13**(2):133-143

[12] Kathiresan K, Bingham BL. Biology of mangroves and mangrove ecosystems. Advances in Marine Biology. 2001;**40**:81-251

[13] Rabinowitz D. Seven forms of rarity. In: Synge H, editor. The Biological Aspects of Rare Plant Conservation. Chichester, UK: Wiley; 1981. pp. 205-217

[14] Rabinowitz D, Cairns S, Dillon T. Seven forms of rarity and their frequency in the flora of the British Isles. In: Soulé ME, editor. Conservation Biology: The Science of

Scarcity and Diversity. Massachusetts: Sinauer Associates; 1986

[15] Bowman WD, Hacker SD, Cain ML. Ecology. Massachusetts: Sinauer Associates; 2017

[16] Wright S. Evolution in Mendelian populations. Genetics. 1931;**16**:97-159

[17] Shaffer ML. Determining minimum viable population sizes for the grizzly bear. Bears: Their Biology and Management. 1983;**5**:133-139

[18] Nash R. The American invention of national parks. American Quarterly. 1970;**22**:726-735

[19] Myers N, Mittermeier RA, Mittermeier CG, Da Fonseca GA, Kent J. Biodiversity hotspots for conservation priorities. Nature. 2000;**403**:853

[20] Numbere AO, Miamaitijiang M. Mapping of nypa palm invasion of mangrove forest using low-cost and high resolution UAV digital imagery in the Niger Delta, Nigeria. Current Trends in Forest Research. 2019;**6**:1-8

[21] McArthur RH, Wilson EO. The Theory of Island Biogeography. New Jersey: Princeton University Press; 2001

Chapter 5

Modeling the Past and Current Distribution and Habitat Suitability for Two Snake-eyed Skinks, *Ablepharus grayanus* and *A. pannonicus* (Sauria: Scincidae)

Rasoul Karamiani, Nasrullah Rastegar-Pouyani and Eskandar Rastegar-Pouyani

Abstract

The study of the climate variability in the past and present, correlating those with changes in the distribution range of species, has attracted considerable research interest. The genus *Ablepharus* consists of 10 recognized species, of which *A. bivittatus*, *A. grayanus*, and *A. pannonicus* are documented from Iran. In the present study, we modeled with MaxEnt the potential distribution areas and determined the suitable habitats in the past [mid-Holocene (MH) and the last interglacial (LIG)] and their current distribution for two species of snake-eyed skinks (*A. grayanus* and *A. pannonicus*) separately. Models of the species indicated good fit by the average high area under the curve (AUC) values (*A. grayanus* = 0.929 ± 0.087 and *A. pannonicus* = 0.979 ± 0.007). Precipitation of the driest quarter of the year, mean temperature of the coldest quarter of the year, and precipitation of the driest month variables made important contributions to *A. grayanus*. Two important climate variables contributed importantly to *A. pannonicus*: temperature seasonality and mean temperature of the wettest quarter of the year and one topographic variable, slope. We conclude that these variables form a natural barrier for species dispersal. The MH and the LIG models indicated a larger suitable area than the current distribution.

Keywords: climate condition, suitable habitat, potential distribution, mid-Holocene, last interglacial

1. Introduction

Climate change plays an important role on the species distributions of biota. The response of species to persistent climate changes may be as follows: (1) consistently in situ at their tolerance limits, (2) changing ranges to regions where climate is within the species tolerance limits, and (3) extinction [1, 2]. During the Pleistocene, several ice sheets in the Northern Hemisphere occurred at intervals of around 40,000–100,000 years [2]. The glaciations were separated by interglacial periods [3].

During interglacial periods, the climate warmed, and forests returned to areas that once supported tundra vegetation [2]. During the last interglacial period (LIG: 150,000–120,000 years), temperature gradient increased in polar regions toward lower latitudes and caused sea level rise and reduction of ice sheets [4]. Briefly, the climate of the last interglacial had a relatively stable warm period [5]. Kerwin et al. [6] simulated terrestrial conditions at the mid-Holocene (6 ka) that indicated summer temperatures were warmer than at present in the high-latitude Northern Hemisphere. But during the mid-Holocene, northern Africa, Arabia, and southern Asia underwent conditions much wetter than at present, these conditions resulting in both African and Asian monsoons [7, 8].

Analyzing species distribution models can help in conservation planning [9] and in understanding theoretical research [10] on ecological and evolutionary processes [1]. Species distribution models can be used to investigate the effect of climate changes on distributions and abundances of species [11], to determine biodiversity [12] and biogeographical patterns [13], to predict potential distribution [14], and to appraise possible future changes in the diversity [15]. Lizards, like other ectotherms [16], provide excellent models for analysis of species distribution under climate change [2]. MaxEnt is a general approach for characterizing probability distributions from small sample sizes [17–19]. MaxEnt estimates the probability distribution of maximum entropy (i.e., closest to uniform) based on environmental variables spread over the survey area [20, 21].

The Scincidae family has more than 25% of all living genera and species of lizards [22]. The genus *Ablepharus* (Fitzinger, 1823) encompasses 10 valid species: *A. bivittatus* (Menetries, 1832), *A. budaki* (Göcmen, Kumlutas & Tosunoglu, 1996), *A. chernovi* (Darevsky, 1953), *A. darvazi* (Jeremčenko & Panfilov, 1990), *A. deserti* (Strauch, 1868), *A. grayanus* (Stoliczka, 1872), *A. kitaibelii* (Bibron & Bory, 1833), *A. lindbergi* (Wettstein, 1960), *A. pannonicus* (Fitzinger, 1824), and *A. rueppellii* (Gray, 1839) which are distributed in Europe, Turkey, Syria to Egypt, Azerbaijan, Armenia, Caucasus, Tajikistan, Kazakhstan, Kyrgyzstan, Uzbekistan, Turkmenistan, Afghanistan, Iran, Iraq, United Arab Emirates, Pakistan, and NW India [23–28]. The genus *Ablepharus* in the molecular phylogenic aspect is a sister taxon of the central and East Asian *Asymblepharus* [29]. *Ablepharus bivittatus* (Menetries, 1832), *A. grayanus* (Stoliczka, 1872), and *A. pannonicus* (Fitzinger, 1824) occur in Iran [30, 31].

Ablepharus grayanus was first described as *Blepharosteres grayanus* from Waggur District, northeast Kutch, India [26]. Later, Fühn [24] regarded it as a subspecies of *A. pannonicus* based on examination of a few specimens (three *A. grayanus*, four *A. pannonicus*). *Ablepharus grayanus* (Stoliczka, 1872) is now regarded as a distinct species. *Ablepharus grayanus* (Stoliczka, 1872) has a distribution range from northern and western India through Pakistan and Afghanistan to Eastern Iran [30, 31]. Researchers based on the morphological characters identified different species and subspecies—*A. brandtii* (Strauch, 1868) from Samarkand, Turkestan; *A. pusillus* (Blanford, 1874) from Basra, Iraq; *A. brandtii* vs. *brevipes* (Nikolsky, 1907) from Dech-i-Diz and Karun River, Iran; *A. persicus* (Nikolsky, 1907) from Shahrud, Iran; and *A. p. pannonicus* and *A. p. grayanus* [24]—in wide distribution range of *A. pannonicus*, that all species regarded to synonym *A. pannonicus* by Anderson [30].

The general aim of this chapter is (1) to identify potential areas of distribution during three periods of the past, last interglacial (LIG: ~120,000–140,000 years BP) and mid-Holocene (MH: ~6000 years BP), (2) to describe current (~1950–2000) distribution and suitable habitat, and to understand the biogeographical patterns of the two mentioned species in Asia.

2. Material and methods

2.1 Study area and records

The study area encompasses the whole Iranian territory. We assembled the species occurrence data for each species based on a systematic biological survey by walking randomly through the habitat from 09:00 to 12:00 AM and 15.00 PM to evening (much of the activity time of species) during spring to summer 2010 and 2015. We used localities mentioned in previous studies (e.g., Anderson [30]; Vyas [28]). *Ablepharus grayanus* specimens were collected, and their distribution data were recorded (34 recorded) from Sistan and Baluchestan and Kerman Provinces, southeastern Iran. We gathered distribution data of *A. pannonicus* specimens collected under rocks or leaves on the floor of oak forest in the Zagros Mountains and in between the meadow grass in the Darvishab River Park (Baghmalek, Khuzestan Province) and recorded the exact location using the global positioning system (GPS). In other areas (Esfahan, Ilam, Kermanshah, Khorasan Razavi, Kurdistan, Lorestan, Mazandaran, Qum, Semnan, Zanjan, and Yasuj Provinces), we observed *A. pannonicus* in between the grasslands, shrubs, and steppes, and exact coordinates were marked with GPS (108 recorded).

2.2 Data set and analysis

We implemented maximum entropy modeling (MaxEnt, 3.3.3e http://www.cs.princeton.edu/~schapire/MaxEnt) of species geographic distributions with default parameters of the data to test samples. We examined 19 bioclimatic variables and 2 topographical variables with grids approximately 1 km^2 precision (30 s × 30 s) for contemporary (~1950–2000) and 10 km^2 precision (5 min × 5 min); we also examined 19 bioclimatic variables in the past (LIG and MH) in the related part of the world (Asia) [32, 33] (www.worldclim.org) (see the Appendix). To identify the correlation ratios between variables and presence records, openModeller (V. 1.0.7) [34] was used. Then we used SPSS IBM (version 22) for Pearson correlation coefficient [17]. We selected variables with a Pearson correlation lower than 0.75 to choose the variables that are ecologically important for species separation according to our observations and to describe habitat [35]. We conducted MaxEnt software with 10 replicates of the analysis that yield the best model for the studied species. MaxEnt provides state distribution models by the receiver operating characteristic (ROC) plots; ROC curves plot true-positive rate against false-positive rate [21, 36]. A value of the area under the curve (AUC) of 0.5–0.7 is taken to indicate that the result is a stochastic prediction [37, 38], values of 0.7–0.9 suggest useful models, and the values more than 0.9 indicate high accuracy [39]. We used DIVA-GIS 7.3.0.1 software for the mean predicted map and a logistic output of present records with suitability ranging from zero (unsuitable habitat) to one (the best suitable habitat) [40].

3. Results

The final models in the present study showed good match and closely fitted the presence of the two species recorded in the study areas, as suggested by high AUC values (*A. grayanus* = 0.929 ± 0.087 and *A. pannonicus* = 0.979 ± 0.007). Moreover, two variables contributed for both species (BIO3 and slope), six variables for *A. grayanus*, and six variables for *A. pannonicus* were detected separately (**Table 1**). The last models in the mid-Holocene simulated high AUC values (*A. grayanus* = 0.975 ± 0.019

and *A. pannonicus* = 0.988 ± 0.006). In addition, three variables were important for both species, one variable for *A. grayanus*, and three variables for *A. pannonicus* were identified separately (**Table 2**). The last interglacial showed high AUC values (*A. grayanus* = 0.975 ± 0.019 and *A. pannonicus* = 0.988 ± 0.006) (**Table 3**). During this time, four variables for *A. grayanus* and six variables for *A. pannonicus* were recognized separately.

The model for *A. grayanus* predicted the distribution range presence of the species in the riparian and wet areas of northwest India, through Pakistan and Afghanistan, and oases and palm groves of the eastern and southeastern Iran. That distribution of the species was verified by using a comparison of environmental variables. Moreover, the climate variable model suggests that there are more suitable potential regions in the United Arab Emirates, Oman, Saudi Arabia, Iraq, Jordan, central Turkey, north Syria, south Turkmenistan and Uzbekistan, and west of China. The MH and the LIG simulated the distribution model for *A. grayanus*

Variable	Description of variables	*A. grayanus*	*A. pannonicus*
BIO2	Annual daily temperature difference (minimal temperature maximal temperature)		0.5
BIO3	Isothermality [(BIO2/BIO7) × 100]	11.4	8.2
BIO4	Temperature seasonality (standard deviation × 100)		27
BIO5	Maximum temperature of the warmest month	1.1	
BIO8	Average temperature of the wettest quarter of the year		18.5
BIO9	Average temperature of the driest quarter of the year	23.3	
BIO11	Average temperature of the coldest quarter of the year		16
BIO14	Precipitation of the driest month	18.4	
BIO15	Seasonality of precipitation (coefficient of variation)		10.5
BIO17	Precipitation of the driest quarter of the year	24	
BIO19	Precipitation of the coldest quarter of the year	15.4	
Slope	Slope	6.5	19.2

Table 1.
Relative of variables (in percentages) at the current period (1950–2000) used in MaxEnt model for the two studied species of the genus Ablepharus.

Variable	Description of variables	*A. grayanus*	*A. pannonicus*
BIO2	Annual daily temperature difference (minimal temperature maximal temperature)	2.1	0.6
BIO3	Isothermality [(BIO2/BIO7) × 100]	22.8	33.9
BIO4	Temperature seasonality (standard deviation × 100)		27.5
BIO7	Temperature annual range (BIO5–BIO6)	59.7	1
BIO8	Average temperature of the wettest quarter of the year		16.6
BIO9	Mean temperature of the driest quarter of the year	15.3	
BIO15	Seasonality of precipitation (coefficient of variation)		20.3

Table 2.
Relative of variables (in percentages) at the mid-Holocene, 6000 years ago (6 ka), used in MaxEnt model for the two studied species of the genus Ablepharus.

Variable	Description of variables	*A. grayanus*	*A. pannonicus*
BIO2	Annual daily temperature difference (minimal temperature maximal temperature)		15.5
BIO3	Isothermality [(BIO2/BIO7) × 100]		28.8
BIO4	Temperature seasonality (standard deviation × 100)		17
BIO7	Temperature annual range (BIO5–BIO6)		7.2
BIO8	Average temperature of the wettest quarter of the year		20.7
BIO9	Average temperature of the driest quarter of the year	8.5	
BIO14	Precipitation of the driest month	56	
BIO15	Seasonality of precipitation (coefficient of variation)		10.7
BIO17	Precipitation of the driest quarter of the year	16.4	
BIO19	Precipitation of the coldest quarter of the year	19.2	

Table 3.
Relative of variables (in percentages) at the last interglacial, 120,000 years ago (120 ka), used in MaxEnt model for two species of the genus Ablepharus.

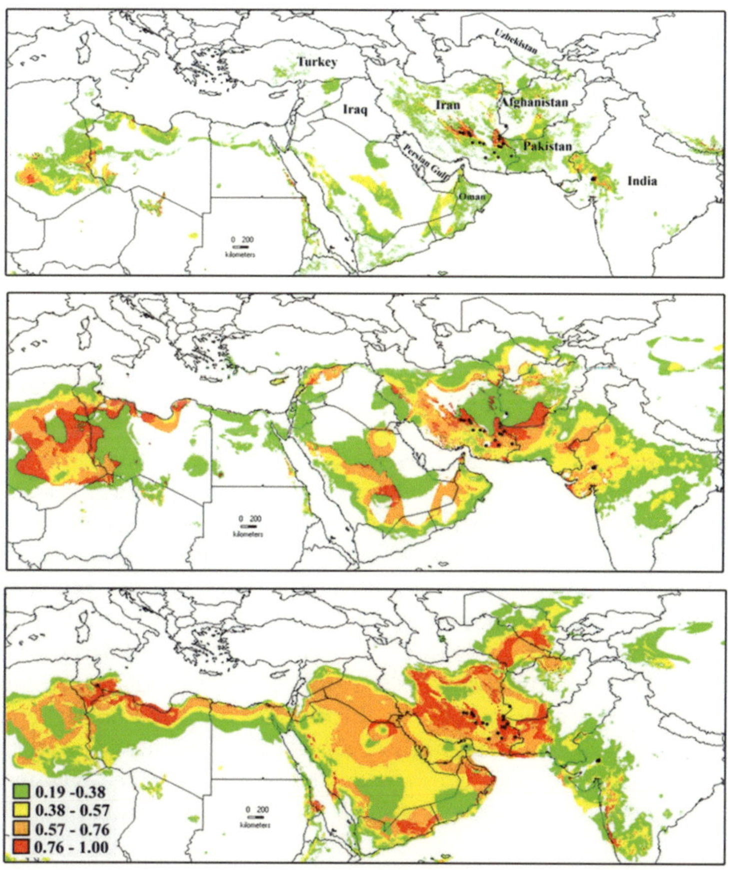

Figure 1.
Distribution map of Ablepharus grayanus in southwestern Asia and much of their potential distribution pattern in the region during: (A) current period (1950–2000); (B) the mid-Holocene, 6000 years ago (6 ka); and (C) the last interglacial, 120,000 years ago (120 ka). The four colored squares on the bottom left indicate the result of stochastic prediction of present species. The black circles refer to the collected specimens.

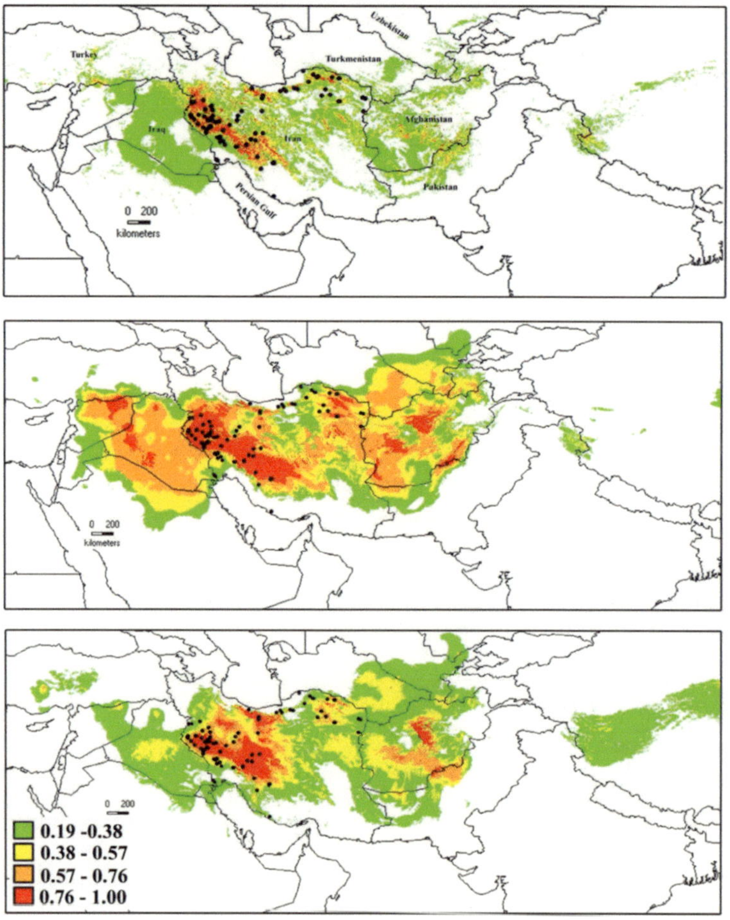

Figure 2.
Distribution map of Ablepharus pannonicus in southwestern Asia and much of their potential distribution pattern in the region during: (A) current period (1950–2000); (B) the mid-Holocene, 6000 years ago (6 ka); and (C) the last interglacial, 120,000 years ago (120 ka). The four colored squares on the bottom left indicate the result of stochastic prediction of present species. The black circles refer to the collected specimens.

that were more suitable areas than present in southwestern Asia today (**Figure 1**). The model for *A. pannonicus* predicted the occurrence of range of the species in steppe areas, grassy, rocky hills separated by oak forest of the Zagros Mountains in the west, and palm groves in southwestern Iran. In addition to the mentioned habitat, the distribution range model of the species predicted that *A. pannonicus* occurs in Iraq, Kuwait, Pakistan, Afghanistan, Tajikistan, Turkmenistan, Uzbekistan, and suitable potential northeast in Syria, Turkey, Kazakhstan, and patchwork areas of northern India. The simulated MH distribution range model for *A. pannonicus* had continuous restriction in east Syria, throughout Iraq, and north Saudi Arabia toward southeastern Turkmenistan. Also, simulated suitable potential fragmented areas of north India and central China were demonstrated. The LIG simulated distribution ranges were the same as the MH suitable potential habitat (**Figure 2**).

4. Discussion

Our results verify the known distribution of the minor snake-eyed skink (*A. grayanus*) and Asian snake-eyed skink (*A. pannonicus*) based on current climatic conditions. The eastern regions of the Iranian Plateau, part of the areas of Afghanistan, northwest India, and Pakistan had the highest suitability for *A. grayanus*, during three time periods (current, MH, LIG). In the eastern Iranian Plateau, *A. grayanus* occurs in the natural parks (e.g., Khobar National Park and the area of the Presidential Museum in Rafsanjan, Kerman Province) and palm graves (Sistan and Baluchestan Province). Recorded from Pakistan at oases, grasslands, backyard gardens, grass fields in the Indus riparian system by Khan [41, 42]. Vyas [28] mentioned three localities (Wagger village of Kutch district, Gujarat; Mount Abu of Sirohi district, Rajasthan; Jessore Wildlife Sanctuary, Gujarat) from India for the species. Model ranges of current distribution predicted areas of western Afghanistan that had conditions suitable as for the same regions mentioned in Pakistan. The model predicted the presence of *A. grayanus* in the United Arab Emirates and Oman but recorded by Gardner [43] as *A. pannonicus*.

Figure 3.
Habitat of Ablepharus pannonicus in Kermanshah, Ilam Provinces, western Iran (A, B), Khuzestan (C), and Fars (D) Provinces, southwestern Iran. The specimens were collected under a relatively small plate stone or under the dead oak leaves, grassland, or steppes. (E, F) Habitat of Ablepharus grayanus in southeastern Iran. The specimens were found under the dead palm leaves and grassland in parks.

The suitable habitats for *A. pannonicus* were in Iran, Pakistan, Afghanistan, and Central Asia (Tajikistan, Turkmenistan, and Uzbekistan). In Iran, *A. pannonicus* was present in the majority of habitat types [30] except deserts, showing the effect of barriers on dispersion of the terrestrial species. This lizard inhabited palm groves (Abadan and Mahshahr), Karoon River shore region, and Darvishab River Park of Khuzestan Province, southwestern Iran (**Figure 3**). It was absent in the steppes of northwestern Iran, probably, due to competition with *A. bivittatus*. Therefore, *A. grayanus* and *A. pannonicus* prefer different climatic conditions across the Middle East and Central Asia. In addition, our results showed that the distributions of these species are restricted by different climatic conditions.

The occurrence and the presence of *A. grayanus* are more influenced by precipitation of the driest quarter of the year (24%), mean temperature of the coldest quarter of the year (23.3%), and precipitation of the driest month (18.45%). Therefore, it is more likely to be found in hot regions under the influence of the rainy monsoon. The prevalence of *A. pannonicus* is more impacted by temperature seasonality (27%), slope (19.2%), and mean temperature of the wettest quarter of the year (18.5%). Due to relationship between temperature and humidity, we claim that seasonal temperatures, especially during the spring, are the most effective factors for suitable habitat.

The models simulated at the MH distribution of *A. grayanus* were highly influenced by precipitation of the driest quarter of the year (59.7%), isothermality (22.8), and mean temperature of the driest quarter of the year (15.3) which resulted from both African and Asian rainy monsoons. Those established damp environments and stable habitats for *A. grayanus*. Another species was highly (79.6%) dependent on temperature (isothermality, temperature seasonality, mean temperature of the wettest quarter of the year, and temperature annual range) that indicated the importance of temperature in range extension for *A. pannonicus*. The models simulated at the LIG distribution of *A. grayanus* was influenced by precipitation of the driest month and the driest quarter of the year (72.7%). *A. pannonicus* (89.2%) was dependent on temperature.

From the last simulation models (6 and 120 thousand years ago), it is clear that in those times wider distribution ranges and areas that are now part of unsuitable habitat, at that time, due to better climatic and environmental conditions influenced by monsoon rainfall, would have been a favorable habitat. Finally, study of the effective bioclimatic variables in a species' distribution over time provides heuristic methods for the management of important habitat by conservation assessments of current habitats and identification of habitat suitability. According to results obtained based on this study, the minor snake-eyed skink, *A. grayanus*, and the Asian snake-eyed skink, *A. pannonicus*, are good indicators for assessing the effects of climatic changes on distribution range of the species over time and for understanding biodiversity patterns in Asia.

5. Conclusion

It is expected that lizards inhabiting open habitats are more susceptible to a predator attack than those inhabiting forest habitats [44], since bushy habitat may provide suitable refuges for lizards. The Asian snake-eyed skink, *Ablepharus pannonicus* (Fitzinger, 1823), was found in the Zagros Mountains among sparse annual grasses, near thorny bushes, natural parks, and under the dead oak leaves. The minor snake-eyed skink *Ablepharus grayanus* (Stoliczka, 1872) lives in palm groves and near rivers in southeastern Iran.

According to results obtained based on this study, the minor snake-eyed skink, *A. grayanus*, and the Asian snake-eyed skink, *A. pannonicus*, are good indicators for assessing the effects of climatic changes on distribution range of the species over time and for understanding biodiversity patterns in Asia.

Acknowledgements

We are grateful to Steven C. Anderson for checking, editing, and improving the manuscript. We thank Hassan Salehi, Mousa Mahmoodi, Hurmuz Nematzadeh, Ali Gholamifard, Sabzali Rasouli, Hiwa Faizi, Mohsen Takesh, Ehsan Damadi, Morteza Akbarpour, and Seyyed Saeed Hosseinian Yousefkhani for assisting us with field-work in Iran. Also we are grateful to Razi University (Kermanshah, Iran) authorities for the financial support during the fieldwork.

A. Appendix

Characters	Definition
Altitude	Altitude
BIO1	Annual mean temperature
BIO2	Mean diurnal range [mean of monthly (max temp–min temp)]
BIO3	Isothermality [(BIO2/BIO7) × 100]
BIO4	Temperature seasonality (standard deviation × 100)
BIO5	Maximum temperature of the warmest month
BIO6	Minimum temperature of the coldest month
BIO7	Temperature annual range (BIO5–BIO6)
BIO8	Mean temperature of the wettest quarter of the year
BIO9	Mean temperature of the driest quarter of the year
BIO10	Mean temperature of the warmest quarter of the year
BIO11	Mean temperature of the coldest quarter of the year
BIO12	Annual precipitation
BIO13	Precipitation of the wettest month
BIO14	Precipitation of the driest month
BIO15	Precipitation seasonality (standard deviation / mean)
BIO16	Precipitation of the wettest quarter of the year
BIO17	Precipitation of the driest quarter of the year
BIO18	Precipitation of the warmest quarter of the year
BIO19	Precipitation of the coldest quarter of the year
Slope	Slope

Table A1.
Climatic variables used to elaborate the models (www.worldclim.org).

Author details

Rasoul Karamiani[1*], Nasrullah Rastegar-Pouyani[1] and Eskandar Rastegar-Pouyani[2]

1 Department of Biology, Faculty of Science, Razi University, Kermanshah, Iran

2 Department of Biology, Faculty of Science, Hakim Sabzevari University, Sabzevar, Iran

*Address all correspondence to: rasoul.karamiani@gmail.com

References

[1] Davis MB, Shaw RG, Etterson JR. Evolutionary responses to changing climate. Ecology. 2005;**86**(7):1704-1714

[2] Sillero N, Carretero MA. Modelling the past and future distribution of contracting species. The Iberian lizard *Podarcis carbonelli* (Squamata: Lacertidae) as a case study. Zoologischer Anzeiger—A Journal of Comparative Zoology. 2013;**252**(3):289-298

[3] Ray LL. The Grate Ice Age. United States Geological Survey General Interest Publications; 1992

[4] Nikolova I, Yin Q, Berger A, Singh UK, Karami M. The last interglacial (Eemian) climate simulated by LOVECLIM and CCSM3. Climate of the Past. 2013;**9**(4):1789-1806

[5] Pickarski N. Vegetation and climate history during the last glacial-interglacial cycle at Lake Van, eastern Anatolia [Thesis]. Bonn: Universitäts- und Landesbibliothek; 2014

[6] Kerwin MW, Overpeck JT, Webb RS, DeVernal A, Rind DH, Healy RJ. The role of oceanic forcing in mid-Holocene northern hemisphere climatic change. Paleoceanography. 1999;**14**(2):200-210

[7] Texier D, De Noblet N, Braconnot P. Sensitivity of the African and Asian monsoons to mid-Holocene insolation and data-inferred surface changes. Journal of Climate. 2000;**13**(1):164-181

[8] Wanner H, Beer J, Bütikofer J, Crowley TJ, Cubasch U, Flückiger J, et al. Mid-to Late Holocene climate change: An overview. Quaternary Science Reviews. 2008;**27**(19):1791-1828

[9] Graham CH, Ferrier S, Huettman F, Moritz C, Peterson AT. New developments in museum-based informatics and applications in biodiversity analysis. Trends in Ecology & Evolution. 2004;**19**(9):497-503

[10] Phillips SJ, Dudík M, Elith J, Graham CH, Lehmann A, Leathwick J, et al. Sample selection bias and presence-only distribution models: Implications for background and pseudo-absence data. Ecological Applications. 2009;**19**(1):181-197

[11] Thomas CD, Cameron A, Green RE, Bakkenes M, Beaumont LJ, Collingham YC, et al. Extinction risk from climate change. Nature. 2004;**427**(6970):145-148

[12] Kleidon A, Mooney HA. A global distribution of biodiversity inferred from climatic constraints: Results from a process-based modelling study. Global Change Biology. 2000;**6**(5):507-523

[13] Yousefkhani SSH, Ficetola GF, Rastegar-Pouyani N, Ananjeva NB, Rastegar-Pouyani E, Masroor R. Environmental suitability and distribution of the Caucasian Rock Agama, *Paralaudakia caucasia* (Sauria: Agamidae) in western and central Asia. Asian Herpetological Research. 2013;**4**(3):207-213

[14] Ananjeva NB, Golynsky EA, Hosseinian YSS, Masroor R. Distribution and environmental suitability of the small scaled rock agama, *Paralaudakia microlepis* (Sauria: Agamidae) in the Iranian Plateau. Asian Herpetology Research. 2014;**5**(3):161-167

[15] Ramirez-Villegas J, Cuesta F, Devenish C, Peralvo M, Jarvis A, Arnillas CA. Using species distributions models for designing conservation strategies of Tropical Andean biodiversity under climate change. Journal for Nature Conservation. 2014;**22**(5):391-404

[16] Barabanov AV, Litvinchuk SN. A new record of the Kurdistan newt (*Neurergus derjugini*) in Iran and potential distribution modeling for the species. Russian Journal of Herpetology. 2015;**22**(2):107-115

[17] Elith J, Graham CH, Anderson RP, Dudík M, Ferrier S, Guisan A, et al. Novel methods improve prediction of species' distributions from occurrence data. Ecography. 2006;**29**(2):129-151

[18] Hernandez PA, Graham CH, Master LL, Albert DL. The effect of sample size and species characteristics on performance of different species distribution modeling methods. Ecography. 2006;**29**(5):773-785

[19] Phillips SJ, Anderson RP, Schapire RE. Maximum entropy modeling of species geographic distributions. Ecological Modelling. 2006;**190**(3):231-259

[20] Pearson RG, Raxworthy CJ, Nakamura M, Townsend Peterson A. Predicting species distributions from small numbers of occurrence records: A test case using cryptic geckos in Madagascar. Journal of Biogeography. 2007;**34**(1):102-117

[21] Kaliontzopoulou A, Brito J, Carretero M, Larbes S, Harris D. Modelling the partially unknown distribution of wall lizards (*Podarcis*) in North Africa: Ecological affinities, potential areas of occurrence, and methodological constraints. Canadian Journal of Zoology. 2008;**86**(9):992-1001

[22] Uetz P, Hošek J. The Reptile Database. 2016. Available from: http://www.reptile-database.org [Accessed: April]

[23] Leviton AE. Report on a Ċollection of Reptiles from Afghanistan. Academy; 1959

[24] Fühn I. Revision and redefinition of the genus *Ablepharus* Lichtenstein, 1823 (Reptilia, Scincidae). Revue Roumaine de Biologie, Série de Biologie Animale. 1969;**14**:23-41

[25] Fühn I. The "Polyphyletic" origin of the genus *Ablepharus* (Reptilia, Scincidae): A case of parallel evolution. Journal of Zoological Systematics and Evolutionary Research. 1969;**7**(1):67-76

[26] Stoliczka F. Notes on the reptilian and amphibian fauna of Kachh. Proceedings of the Asiatic Society of Bengal. 1872;(5):71-85

[27] Khan MS. Key and checklist to the lizards of Pakistan. Herpetozoa. 2002;**15**(3/4):99-119

[28] Vyas R. Preliminary survey on reptiles of Jassore Wildlife Sanctuary, Gujarat State, India. Russian Journal of Herpetology. 2011;**18**(3):210-214

[29] Pyron RA, Burbrink FT, Wiens JJ. A phylogeny and revised classification of *Squamata*, including 4161 species of lizards and snakes. BMC Evolutionary Biology. 2013;**13**(93):1-53

[30] Anderson SC. The Lizards of Iran. Society for the study of Amphibians and Reptiles. Contributions to Herpetology. Vol. 15. Ohio: Oxford; 1999. pp. I-VII, 1-442

[31] Karamiani R, Rastegar-Pouyani N, Rastegar-Pouyani E, Akbarpour M, Damadi E. Verification of the Minor Snake-eyed Skink, *Ablepharus grayanus* (Stoliczka, 1872) (Sauria: Scincidae), from Iran. Zoology in the Middle East. 2015;**61**(3):226-230

[32] Hijmans RJ, Cameron SE, Parra JL, Jones PG, Jarvis A. Very high resolution interpolated climate surfaces for global land areas. International Journal of Climatology. 2005;**25**(15):1965-1978

[33] Otto-Bliesner BL, Marshall SJ, Overpeck JT, Miller GH, Hu A. Simulating Arctic climate warmth and

icefield retreat in the last interglaciation. Science. 2006;**311**(5768):1751-1753

[34] de Souza Muñoz ME, De Giovanni R, de Siqueira MF, Sutton T, Brewer P, Pereira RS, et al. openModeller: A generic approach to species' potential distribution modelling. GeoInformatica. 2011;**15**(1):111-135

[35] Yousefkhani SSH, Tehrani SJ, Moodi B, Gül S. Distribution patterns and habitat suitability for three species of the genus Hyla Laurenti, 1768 in the Western Palearctic. Turkish Journal of Zoology. 2016;**40**(2):257-261

[36] Phillips SJ, Dudík M, Schapire RE. A maximum entropy approach to species distribution modeling. In: Proceedings of the Twenty-First International Conference on Machine Learning; ACM; 2004. p. 83

[37] Raes N, ter Steege H. A null-model for significance testing of presence-only species distribution models. Ecography. 2007;**30**(5):727-736

[38] Gallien L, Douzet R, Pratte S, Zimmermann NE, Thuiller W. Invasive species distribution models—How violating the equilibrium assumption can create new insights. Global Ecology and Biogeography. 2012;**21**(11):1126-1136

[39] Manel S, Williams HC, Ormerod SJ. Evaluating presence–absence models in ecology: The need to account for prevalence. Journal of Applied Ecology. 2001;**38**(5):921-931

[40] Hijmans R, Cruz M, Rojas E, Guarino L. DIVA-GIS, version 1.4. A geographic information system for the management and analysis of genetic resources data. Plant Genetic Resources Newsletter. 2001;**127**:15-19

[41] Khan MS. Herpetology of habitat types of Pakistan. Pakistan Journal of Zoology. 1999;**31**(3):275-289

[42] Khan MS. Herpetological Laboratory, Series. Pakistan Journal of Zoology Supplement Series. 2012;**11**:1-12

[43] Gardner A. Mapping the terrestrial reptile distributions in Oman and the United Arab Emirates. ZooKeys. 2009;**31**:165

[44] Schoener TW, Schoener A. Ecological and demographic correlates of injury rates in some Bahamian *Anolis* lizards. Copeia. 1980;**1980**:839-850

Chapter 6

Influence of Floods on Spatial Variability of Wetslums Using Geo-information Techniques: A Case Study of a Specific Human Habitat in Korail, Dhaka

Koen Olthuis, Kasirajan Mahalingam, Pierre-Baptiste Tartas and Chris Zevenbergen

Abstract

A previous study by the authors found that most slum physical upgrading projects traditionally focus on basic services, elaborated with citywide data, without addressing locational and environmental aspects. However, such data tends to hide the highly heterogeneous nature of slums. Thus, this paper's objective is to further investigate the spatial variation resulting from the influence of water on the living conditions within slums, by proposing a framework that can quantify it. This framework is used to establish the correlation between the impact of flooding denoted by a Flood Proneness Index and living conditions denoted by a Slum Living Conditions Index in Korail, Dhaka. The paper concludes that in Korail, both flooding and living conditions exhibited spatial variability, with the former seeming to have a significant influence on the latter, particularly in areas located close to or on a water body. As a result, the paper proposes to define slums which exhibit considerable correlation between flooding and living conditions as wetslums. The analysis in Korail further revealed cluster formation and as such strengthens the hypothesis of locational variability in these specific human habitats. Subsequently, wetslum areas that require resilient physical upgrading are identified thus highlighting the importance for location-specific upgrading.

Keywords: flood proneness, living conditions, slum index, upgrading, slum heterogeneity

1. Introduction

A study made by Olthuis et al. in 2015 [1] investigated slum upgrading projects and revealed that most physical upgrading projects have focused primarily on improving household-based statistics such as water and sanitation while overlooking environmental and locational adaptation. This study also showed how location impacts diversity and influences evolution patterns in slums. As environmental and locational factors are crucial to the living conditions of a slum [2], ignoring

IntechOpen

these aspects is a threat to the effectiveness of any slum physical upgrading project. According to Olthuis et al. [1], the disregard for locational aspects may be attributed to the dominance of the UN criteria for defining slums. These criteria force authorities to develop numerical datasets that include characteristics such as number of households with access to water, sanitation, etc.

In addition, previous research has started to investigate the heterogeneity of slums by, for instance, exposing how aggregating data at city-level scale hides the spatial variation of slums [3]. Further studies [4, 5] have highlighted the considerable spatial variability in the extent to which a slum settlement exhibits characteristics of a slum—the so-called slumness [4]—and have discussed how slums are heterogeneous human habitats in all aspects ranging from socioeconomic to vulnerability [5]. These researches also have started to expose the dichotomous nature of existing slum classification systems that ignores their physical and social diversity. Thus, spatial variation or heterogeneity is an important but overlooked aspect of slums and has been studied only to a very limited extent in the past [6].

As a result, this paper intends to present a step toward filling this gap, through the study of the influence of location on the living conditions of slums near or on water. One of the influences of water on slums is the issue of flooding. Numerous examples of flooding in slums from Nigeria to Peru have been recorded and studied [5, 7–15]. The breadth of literature on the impact of flooding on slums confirms their vulnerability to such natural disasters, validating its choice as a major locational influence to be studied. Hence, this paper aims to further conceptualize the degree to which a slum settlement is spatially variable and subsequently addresses the influence of water and its flooding characteristics in the given variability.

Therefore, the end goal of this paper is to propose a framework that can quantify, in a studied slum, the influence of water—the proximity to a water body and flood risk vulnerability faced by its inhabitants—on its living conditions. To do so, the integral nature of two factors in defining the vulnerability of slums located near or on the water—flooding susceptibility and living conditions—needs to be emphasized. To this end, two indexes will be developed—a Flood Proneness Index and a Slum Living Conditions Index—taking Korail in Dhaka (Bangladesh) as a case study to test this methodology. At the same time, structurally and characteristically similar units will be defined and tested for correlation. Then, the correlation between flooding and living conditions should give us significant insights into the functionality of a slum—a key factor to consider for future physical upgrading projects.

2. Literature review

This section provides an overview of firstly living conditions within slums, secondly the impact of flooding on slums, and lastly the spatial heterogeneity of slums.

2.1 Living conditions in slums

Living conditions in slums have been defined in a range of ways. Gulyani and Bassett [2] have established the "living conditions diamond" as a framework to study the physical aspects of a slum's living conditions. The diamond has four physical components—tenure, infrastructure, unit quality, as well as neighborhood and location—seen as essential to determine the living conditions in slums.

Tenure, or a lack thereof, often is a key precondition determining investment in physical upgrading projects. Infrastructure, including physical services such as water supply or electricity, as well as public services, makes settlements and housing functional. Slums built on or extending into water bodies present a special case. They extensively use water as a major access network. Examples can be found in slums such as Makoko, Nigeria [16], or Isla Verde in Davao City, Philippines [17].

The third physical component—housing unit—investigates the quality of housing. This vertex incorporates building materials and the density of occupancy. The nature of materials used for roofing, foundations, and exterior walls can help determine the building quality. In some cases, slums might be erected on stilts like in Palembang, South Sumatra [14]; Korail, Dhaka [15]; or Ribeira Azul, Brazil [18]. Studies conducted by Flores-Fernandez [19] have showcased the creative approach undertaken by slum dwellers in risky areas to expand and/or develop their settlement. The key findings of this study have elaborated the organic and unique forms of the slum settlement, adapted to the morphological territorial characteristics (such as slope, height, profile, etc.).

Neighborhood and location comprise the final component. The settlements' location and connectivity can indicate how physically and environmentally vulnerable a slum is. Density, physical layout, and the presence or absence of amenities and services such as schools, open spaces, and community facilities are further factors that influence the living conditions of slums. Makoko in Lagos presents an example of a slum where the remote location in the lagoon alongside low status of the inhabitants is believed to lead to serious environmental and infrastructural deficiencies. For instance, this slum has an inadequate access to education and healthcare [16].

Taken together, all four components determine the physical living conditions in slums. In that way, the diamond is a strong framework to study a slum's living conditions. However, data availability for the individual components is often nonexistent or highly dynamic [2]. Any framework attempting to quantify the living conditions in slums therefore will have to be adaptable to changing conditions of slums. However, the lack of up-to-date data does not allow the precise overview of each one of these four components. Yet, modular components of the framework in themselves help understand the living conditions of the slum habitat.

2.2 Flooding in slums

Slums, by definition, are inclined to be located in hazardous areas [2, 9]. These include areas prone to flooding such as river floodplains, foreshore areas on mangrove swamps, or tidal flats [9]. Moreover, the high population density added to a lack of protection against climate change and sea level rise makes the urban poor increasingly vulnerable to flooding [20]. According to the United Nations International Strategy for Disaster Reduction (UNISDR) [21], rare big flooding events account for the biggest losses of lives and assets in urban poor settlements, where smaller more frequent events result in fewer deaths. However, the latter have predominant impact on urban poor due to the fact that there are events affecting their daily life, causing damage to housing, infrastructure, livelihoods, as well as their health. Indeed, the reports from the UNISDR [21] found evidences from cities in Africa, Asia, and Latin America suggesting that the increase in reports of weather-related disasters is a sign of the expansion of informal human habitats.

Douglas et al.'s [9] study of the urban poor in Africa presents the effects of flooding on slums. Among different causes of flooding, the study highlights how slums in urban areas are most often subject to localized flooding events and flooding from small streams. Moreover, the study also specifies that slums located near major rivers and on coastal areas face an additional level of vulnerability and threat.

The main threat caused by floods is not the flood itself but the stagnant water added to water pollution, in other words, prolonged floods [22]. They are caused by extensive urbanization, waterlogging, overly saturated grounds, and blockage of the sewage and drain systems by solid wastes [23]. Indeed, slums' growing population increases waste productions which are accumulating on-site due to the absence of proper waste management. For example, in Dhaka, only half of the total wastes are collected, and no waste collection is made in slums due to access difficulties.

The combination of solid wastes and human ones creates a major threat to slums—human waste could lead to sanitary disaster, and solid ones, when carried by water flow, are partially responsible for the destructions during floods as well as for the blockage of drains and sewages [24, 25]—worsening the risks faced by slum dwellers during prolonged floods. Another dramatic issue is the massive spread of diseases—long-lasting floods generated several health hazards, going from drinkable water contamination to mosquito infestation—and this risk increases in accordance with flood duration which could last several months [26].

The impact of prolonged floods on dwellers drastically limits access to basic needs such as food, drinkable water, medicines, and cloth as well as access to sanitation, shelters, and dry places to sleep [27]. Flooding also disrupts small-scale activities like petty and artisanal trading, thus threatening slum dwellers' livelihoods. Indeed, Kanke Arachchilage [28] shows how flooding disrupts the economic activities of rickshaw pullers. As streets are converted to streams, they are unable to work. Rather than get to a safer area, slum dwellers are often in a state of forced inertia during flooding conditions in order to not displace their assets and social and livelihood networks. Nevertheless, post-disaster, more than 50% of households have to be rebuilt or repaired [15].

2.3 Spatial heterogeneity of slums

Recent literature is increasingly studying the spatial variability apparent in slums and how to measure it [2, 4, 5]. For instance, in a study made in Union Territory, Chandigarh, Rao and Thakur [29] found that 15.5% of slum dwellers lived in high vulnerable areas, 44% in medium and 40.5% in low vulnerable areas.

The slum index introduced by Weeks et al. [4] presents an attempt to measure spatial variability of a slum in Accra. Each housing unit is scored, according to a binary system, as follows:

- If the housing unit does not have piped water, then $slum_1 = 1$ (else 0).
- If there is no toilet and no sewage connection, then $slum_2 = 1$ (else 0).
- If the number of persons per room is greater than 2, then $slum_3 = 1$ (else 0).
- If the building material is less durable, then $slum_4 = 1$ (else 0).
- If the resident is not the owner, then $slum_5 = 1$ (else 0)
- Slum index for each housing unit $(S_h) = \sum (slum_1 \ldots slum_5)$.

The results of the study indicated considerable variability in the "slumness" of the neighborhoods in Accra. Rather than defining households as "slum" or "not slum," the measure adds the number of slum conditions for each housing unit in a defined area to then calculate an average score for each neighborhood. As a result, each slum is placed along a continuum. A later study on the slums of Accra highlights how vulnerability even varies spatially in a single slum settlement [30]. Indeed, Jamestown in Accra, often considered as one slum, is demonstrated as a highly complex place with specific vulnerabilities varying within the neighborhood itself. By quantifying the spatial variability of vulnerability within slums, the slum index could thus help to develop a broader rating and monitoring systems for slums.

Therefore, once structurally similar units—the so-called clusters—are identified through an index, their interdependence can be investigated through spatial autocorrelation [31]. Spatial autocorrelation has moreover been widely utilized to assess the spatial dependency of space by measuring the variables of landscapes that influence spatial variability [32].

3. Framework and methodology

To investigate the relation between locational characteristics and living conditions in slums near water, three stages have been undertaken in the conceptual framework, namely, the creation of a Flood Proneness Index, Slum Living Conditions Index, and spatial pattern analysis (**Figure 1**). These stages are built around data from 2006 to 2016 of Korail, Dhaka, serving as case study in this work.

3.1 Case study: Korail, Dhaka

Surrounded by Banani Lake on the eastern and southern sides, Korail is the largest informal settlement in Dhaka and located in a low-lying, flood-prone area. It started to develop during the late 1980s on vacant high grounds but later expanded into more hazardous low-lying areas with houses built on the flood-prone water edges. Korail is mostly inhabited by people engaged in service jobs in the high-end area on the eastern bank of the lake. Population estimates vary enormously, from 100,000 people [15] living on approximately 90 acres to more recent newspaper reports suggesting numbers as high as 175,000 people living on 170 acres [33], which would point toward an increase of 75,000 people and 80 acres in only 4 years.

The slum's existing living conditions—stemming from its location and high population density—is further exacerbated by unmanaged waste disposal and changing climate and weather conditions. Almost every year, Korail is subjected to extreme and hazardous conditions, due to excessive rainfall, increased heat, and flooding [15, 34]. There have been at least 10 heavy floods that were recorded in Dhaka city between the years from 1954 and 2007 [35].

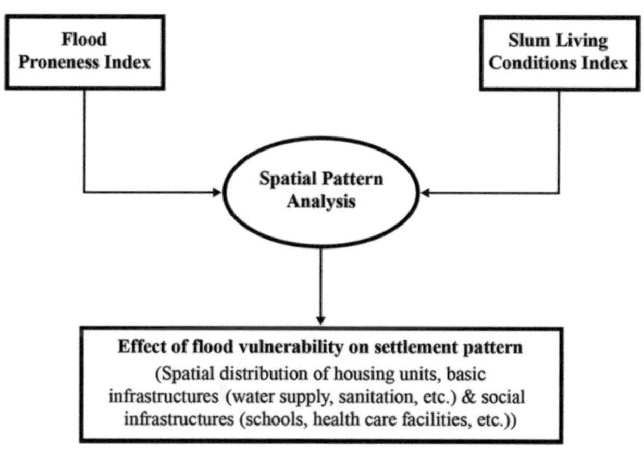

Figure 1.
Conceptual framework to assess flood proneness alongside living conditions [by authors].

Due to the lack of tenure in Korail, both the inhabitants and the government are reluctant to improve their living conditions [15]. Basic facilities, schools, and healthcare facilities are run by NGOs. Cameron [36] highlights the crucial role that NGO primary schools play in Korail, some of which host as much as 600 kids. Other types of education centers are kindergartens that accommodate around 300 kids.

To test the methodology of this study, a statistical analysis in the form of spatial autocorrelation is performed to identify building units that are exhibiting significant correlation. Secondly, these clusters make it possible to highlight spatial variability—through a multi-distance spatial cluster analysis (see Appendix A.1)—within the slum, assess the overall influence that water has on the slum, and thus help place Korail on a spectrum. The slum is situated amid poor environmental condition, siting mostly on low-lying areas which cause waterlogging and flooding in most building units inside the slum [37]. In these aspects, Korail is similar to many other slums around the work, thus making it a suitable case study.

Due to a lack of updated credible data, the conceptual framework (**Figure 1**) has been adjusted for Korail's spatial pattern analysis (**Figure 2**). Thus, data for only two of the four physical components of the living conditions diamond, infrastructure and neighborhood and location, were available and integrated in the framework.

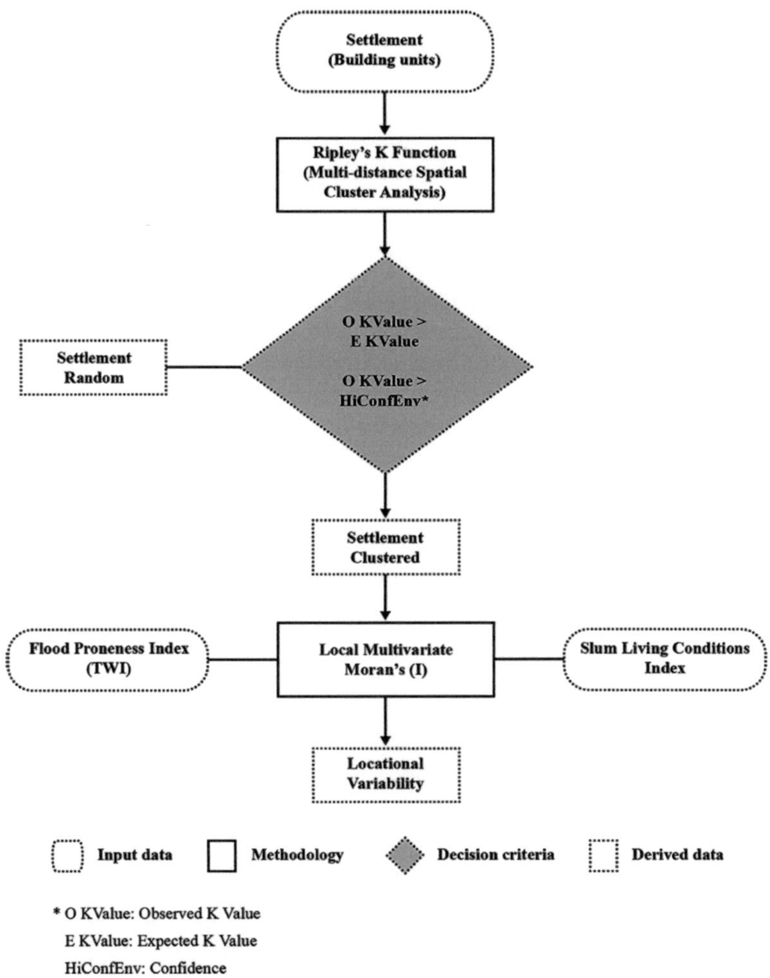

Figure 2.
Spatial pattern analysis in Korail, Dhaka [by authors].

These data are secondary data collected from research papers [15, 35, 37] and conference [7] as well as from institutions and NGO documents [28, 34, 36, 38, 39].

3.2 Stage 1: development of a Flood Proneness Index

A flood index can be determined by looking one or a combination of the following hydrological characteristics.

3.2.1 Physical conditions of the locations

Topographic Wetness Index (TWI): a hydrological analysis can be performed to retrieve a "Topographic Wetness Index" from Digitally Elevated Maps (DEM) to identify areas that are more susceptible to intense flooding due to the location's "slope" and "flow direction."

3.2.2 Dynamic variables

Rainfall data: crucial information required for identifying areas prone to flash floods. A study of rainfall and its intensity can aid the understanding of annual or monthly patterns [40].

Flood inundation: modeling inundation can predict the extent to which a floodplain is at the risk of flooding. Conventional inundation simulation involves overlaying water depths at different cross sections onto a DEM. Alternatively, inundation extents can be interpolated at cross sections [41].

3.2.3 Mapping flood-prone areas through the topographic wetness index in Korail, Dhaka

Due to the limited availability of flood inundation and rainfall data for Korail, open-source elevation data was chosen as the main input for the mapping of flood impacts. The elevation data helps to examine the TWI of Korail. A high indicator is considered to be illustrative for areas which are prone to drain due to excess flows of water during flooding conditions. A hydrological analysis was performed on the basis of TWI data from DEM. Based on this analysis, Korail was mapped into vulnerable and non-vulnerable zones.

The analysis of TWI required minimal data in the form of a DEM, which was obtained from the United States Geological Survey (USGS). An ASTER geo-dataset was retrieved in DEM format (ASTGTM2_N23E090) with the coordinates for Korail, Dhaka. The TWI was retrieved through the software TauDEM [42]. The topographic or compound wetness index is retrieved through the calculation of the ratio between slope and the specific catchment or contributing area [40]. In other words, the wetness index is a slope over area ratio. Although slope in this case is a much complex variable, it is achieved by studying the "flow direction" of steepest slopes in the elevation profile. Subsequently, the corresponding catchment area is identified at the scale of grid cells as "contributing area." Previous studies have considered TWI to be a cost-efficient method to summarize overland flash flood-prone areas [41]. TWI is dependent on two variables—D-infinity flow direction and D-infinity contributing area.

Calculating D-Infinity Flow Direction:

Flow directions are assigned based on the D-infinity flow method, which primarily captures the steepest slope on a planar block-centered grid. The grid presents the elevation value taken to represent the elevation of the center of the corresponding grid cell. The flow directions are recorded as angles in radians and are calculated counterclockwise. The direction of the steepest downward slope is gathered upon the triangular facet of a 3 × 3 grid cell; the resulting flow is the proportion between these two neighboring cells [43]. As seen in **Figure 3**, the grid is further divided into eight triangular facets between each cell and its neighbor.

The downslope vector within each of these triangular facets is taken into account, through which the slope and flow direction associated with the grid cells are produced. Slope is measured as drop/distance which is tan of the slope angle. Downslope vector ranges within or outside 45° angles.

Calculating D-Infinity Contributing Area

Contributing area in this case pertains to a specific catchment area, which is calculated as area per contour length using multiple flow directions or D-infinity method. Contribution (catchment) at each grid cell corresponds to its very own grid length. Equation (1) denotes the execution of TWI, wherein α is the contributing area and $tan\beta$ is the surface slope:

$$TWI = \ln\left(\frac{\alpha}{tan\beta}\right) \quad (1)$$

3.3 Stage 2: development of a Slum Living Conditions Index

Similarly to the hydrological analysis, data on service provision and access to healthcare and education in Korail was limited. Therefore, secondary data such as NGO presence in the slum was used to determine the access to education and health for the Slum Living Conditions Index. The slum was then broken down into building units with access to both education and health centers, neither education nor health centers or only education or health centers, and on or close to water.

The literature review showed that education centers serve roughly 900 settlements within Korail [36]. Due to a lack of concrete data on the service capacity of health centers within the slum, the national percentage for access to primary health center in Bangladesh was taken as an indicator for healthcare provision.

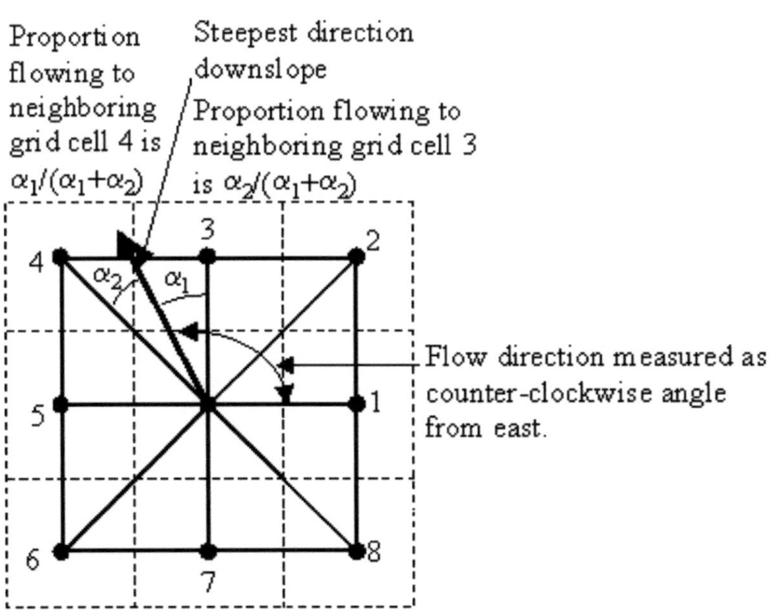

Figure 3.
D-infinity flow direction [43].

According to the Asian Development Bank [38], the index currently stands at 33.4%. As a result, 33% of the settlements within the slum were considered to have access to healthcare provision of static health centers. The location of the education centers and static health centers, which are functioning as general practitioners in the slum, was identified on the basis of BRAC's "frugal maps" [39]. Then, buffer zones were created to delineate building units that fell under immediate access to the said centers, based on the capacity of health or education centers (see Appendix A.2).

In addition, the building density of these units was also included in the Slum Living Conditions Index. Moreover, this factor was also used for a secondary study investigating the repartition of dense building units in the slum. To do so, a temporal study was conducted to assess the density of settlement with respect to distance from the water body (lake). The temporal aspect was in the form of assessment of built units prior to 2011 and after 2011 (till 2016) and based on satellite observation. Then, in GIS environment, built units were converted into point features. These point features were used as input for the point density function, which in turn allowed calculation of density for each point feature (built unit). Similarly, near function was used to calculate the proximity of each point feature from the lake.

As a result, the Slum Living Conditions Index combined service provision, density, and proximity to the water body. In the present case and as mentioned above, the available data were limited. Thus, in order to produce a precise vision of a living conditions, access to primary and updated data for all the four physical components of the 'living conditions diamond' will be necessary. Moreover, data about all the different types of services provided in the studied slum such as water pipe connection, sanitation access, or waste removal, for instance, will also have to be integrated, when existing, in this index.

3.4 Stage 3: analysis of spatial patterns

Then, the relationship between the Flood Proneness Index and the Slum Living Conditions Index is examined. The spatial autocorrelation technique allowed the study to conceive the statistical significance between the said indexes and identify the underlying cluster formations based on these factors. This spatial autocorrelation was performed through GeoDa [44]. Then, in a second time, this software was also used to investigate the influence of proximity to water body (distance) on density of the settlement through the performing of local Moran's I (bivariate) [45].

Through spatial analysis it is possible to identify clusters that represent the varying influences of both flooding and living conditions as well as the influence of proximity of water body to the density of the building units. Thus, two types of clusters (**Figure 4**) and two types of spatial outliers (**Figure 5**) can be identified.

- The two types of clusters include:

    - **High-High (HH)** clusters correspond to areas that are highly prone to flooding and have a high deficiency in living conditions, e.g., a low Slum Living Condition Index, while in the second analysis, these *High-High (HH)* clusters represent high distance from the water body and high density of the building unit.

    - **Low-Low (LL)** clusters correspond to areas that are not very prone to flooding and have a low deficiency in living conditions, e.g., a high Slum Living Conditions Index, while in the second analysis, these *Low-Low (LL)* clusters represent low distance from the water body and low density of the building unit.

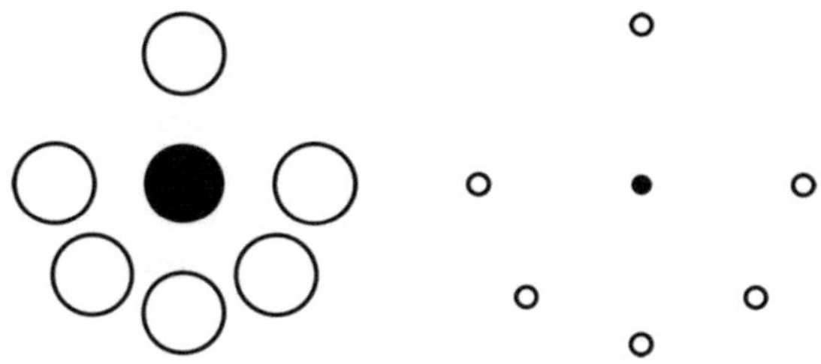

Figure 4.
Spatial clusters. High-High (HH) spatial clusters (left) and Low-Low (LL) spatial clusters (right) (adapted from [45]).

- The two types of spatial outliers include:
    - **High-Low (HL)** outliers correspond to areas that are highly prone to flooding but have low deficiency in living conditions, e.g., a high Slum Living Conditions Index, while in the second analysis, these *high-Low (HL)* outliers represent high distance from the water body and low density of the building unit.
    - **Low-High (LH)** outliers correspond to areas that are not prone to flooding but have a high deficiency in living conditions, e.g., a low Slum Living Conditions Index, while in the second analysis, these *Low-High (LH)* outliers represent low distance from the water body and high density of the building unit.

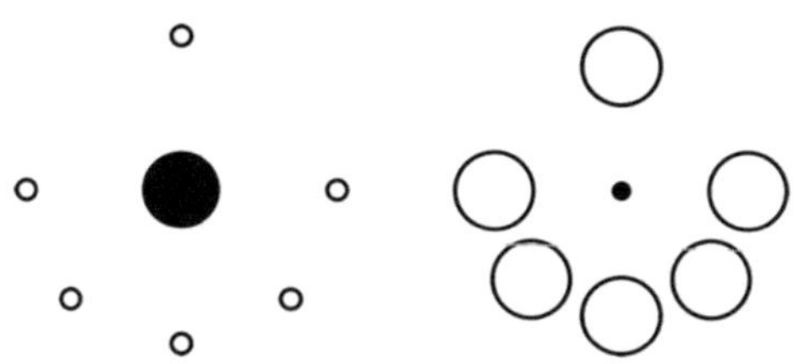

Figure 5.
Spatial outliers. High-Low (HL) spatial outliers (left) and Low-High (LH) spatial outliers (right) (adapted from [45]).

By looking at the combined influence of flooding and living conditions, the clusters aid an understanding between the two variables and help to classify a large slum into specific zones. At the same time, the temporal analysis allows the comprehension of the relation between the density of a settlement and its distance from a water body as well as observing where the newcomers settle. In other words, determine if the slum expands toward the water body.

4. Key findings

Now that the indexes are built, the spatial pattern analysis of Korail is used to produce several maps. These maps will now be utilized to investigate the influence of water on slums' living conditions.

4.1 Main findings

A number of observations can be made from the maps produced with the TWI, the Slum Living Conditions Index, and the spatial pattern analysis.

Firstly, there is a clear spatial delineation between planned and unplanned settlements. **Figure 6** shows that slum settlements have developed, on a majority, in the vulnerable zones—irrespectively of their proximity to the water. In contrast, the non-vulnerable zones are dominated by planned settlements.

Therefore, areas considered unsafe and thus devoid of any planned settlements seem to attract slum dwellers, which correspond to the results obtained by Douglas et al. [9] as well as Gulyani and Bassett [2]. In addition, **Figure 6** also exhibits the presence of building units located close to or on water, which is also demonstrated by **Figure 7**.

This confirms the observation made by Olthuis et al. [1]. Indeed, according to this study, these building units should be an indicator of Korail expansion toward Banani Lake caused by the increase in population occurring in this slum. This would demonstrate that new dwellers arriving in the slum tend to go to even riskier areas. However, this will have to be verified with the temporal study.

At the same time, it would mean that areas with a lower TWI—which are more elevated and thus "safer"—are most likely sought-after destinations in the slum, forcing those dwellers that do not find space here onto more hazardous locations.

This also explains the utilization of construction on stilt or artificial elevation observed in Korail by different research [15, 28], thus illustrating Flores-Fernandez's findings about the creative approach undertaken by slum dwellers to settle in risky areas [19]. As a result, a clear delineation in Korail's development pattern caused by topographic factors, such as elevation, seems to exist.

Secondly, according to **Figure 7**, the settlements which are the closest to Banani Lake exhibit the higher deficiency in living conditions. These observations are understandable due to the fact that these areas are the most exposed to flooding. In addition, the same figure shows that several of these zones correspond to a higher density in building arrangement. This however will have to be confirmed by the temporal study.

Moreover, when considering the maps used to build the Slum Living Conditions Index (see Appendix A.2), it appears that these areas are, for the great majority, not included in the health and educational centers' buffer zones. Nevertheless, when

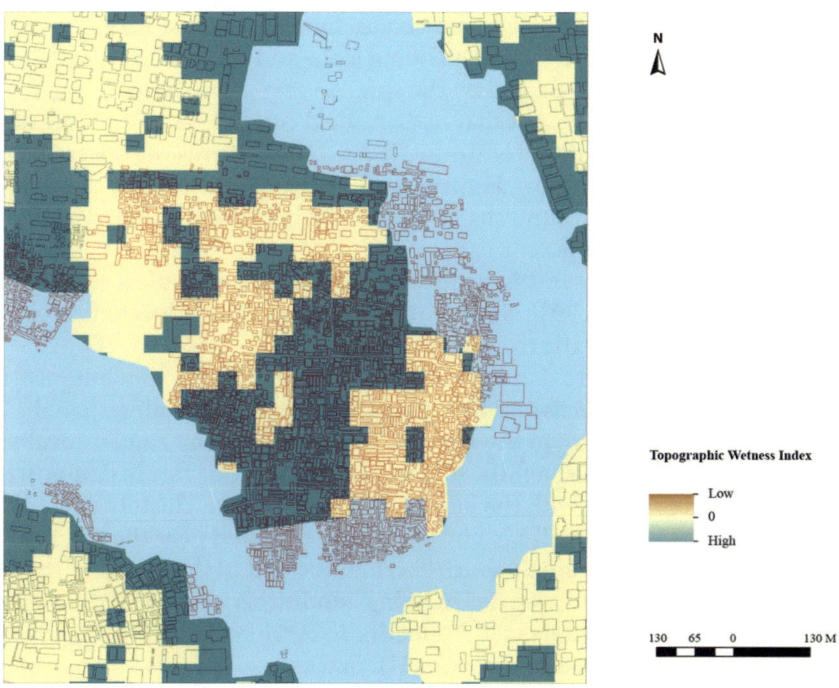

Figure 6.
Mapping flood-prone areas in Korail with TWI [by authors].

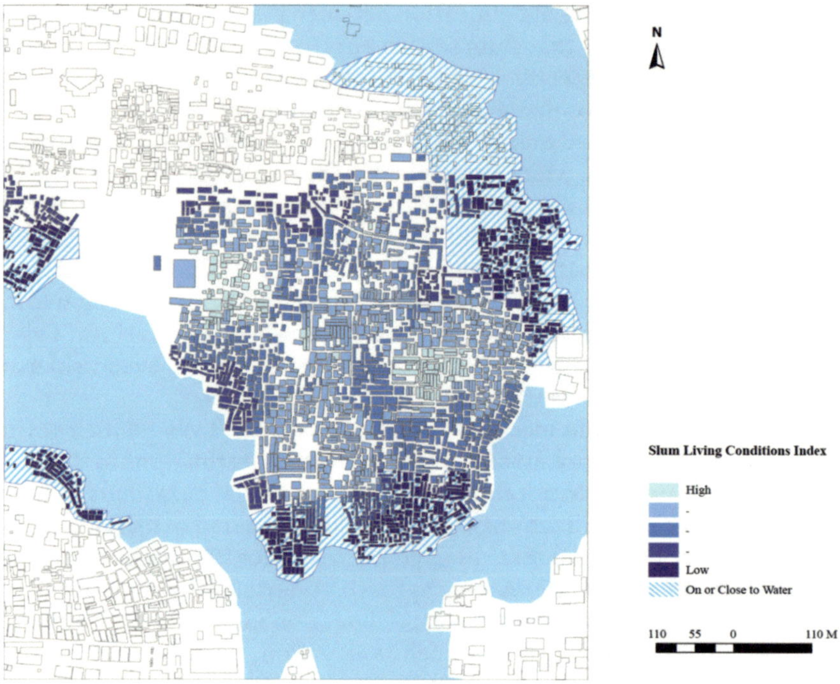

Figure 7.
Mapping the living conditions in Korail with the Slum Living Conditions Index [by authors].

comparing these figures to **Figure 7**, it comes into sight that some areas in close proximity of social centers are under-serviced. This can be attributed to the surge in unavailability of services within the slum [15, 38].

At this point, it seems that flood risks as well as proximity to water greatly affect slum dwellers' living conditions. This observation is similar to the one made by Kanke Arachchilage [28] which demonstrates the reduction of services and the increase of vulnerability for the dwellers living close and on Banani Lake. However, this observation has to be confirmed by the spatial variability study.

Thirdly, **Figure 8** demonstrates that hydrological conditions have a significant influence on the living conditions of Korail. Indeed, it confirms the observations that can be made when comparing **Figures 6** and **7**, where the areas located in the most vulnerable parts, i.e., around the fringes of the lake or on it, have the highest deficiency in living conditions.

In addition, **Figure 8** shows this correlation between hydrological conditions and living conditions with a 99% confidence level. Thus, slums located close to a water body faced issues specific to this localization.

Lastly, **Figure 9** highlights that Korail is not a homogeneous slum but instead presents variable flood patterns and living conditions, thus revealing spatial variability of the vulnerability in the slum [4, 5] and presenting similar results as the ones observed by Jankowska [30]. As a matter of fact, this figure demonstrates the presence of "safe" zones with the identification of two **LL** clusters while also presenting the presence of small **LH** outliers that are located near the Banani Lake.

At the same time, a large north-south corridor located in the center of the slum presents one of the lowest deficiencies in living conditions and concentrates most of the services and social centers (see Appendix A.2) while being located in an area highly prone to flooding—in other words a **HL** outlier.

Furthermore, this map also confirms the observation made with **Figure 7** and by Kanke Arachchilage [28]. Indeed, the clusters located near or on the Banani Lake

DOI: http://dx.doi.org/10.5772/intechopen.85649

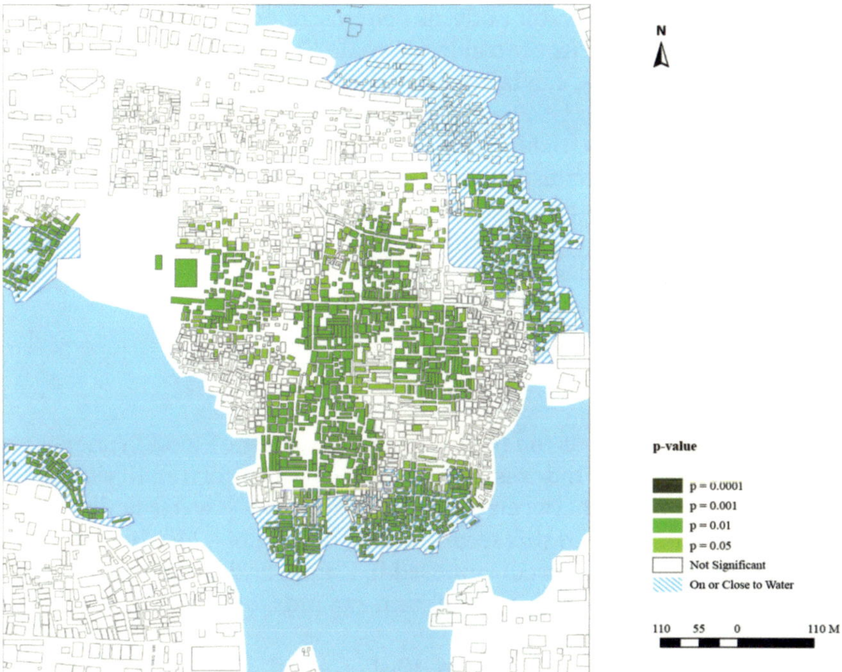

Figure 8.
Significance map (local Moran's I) of the correlation between flood proneness and living conditions in Korail [by authors].

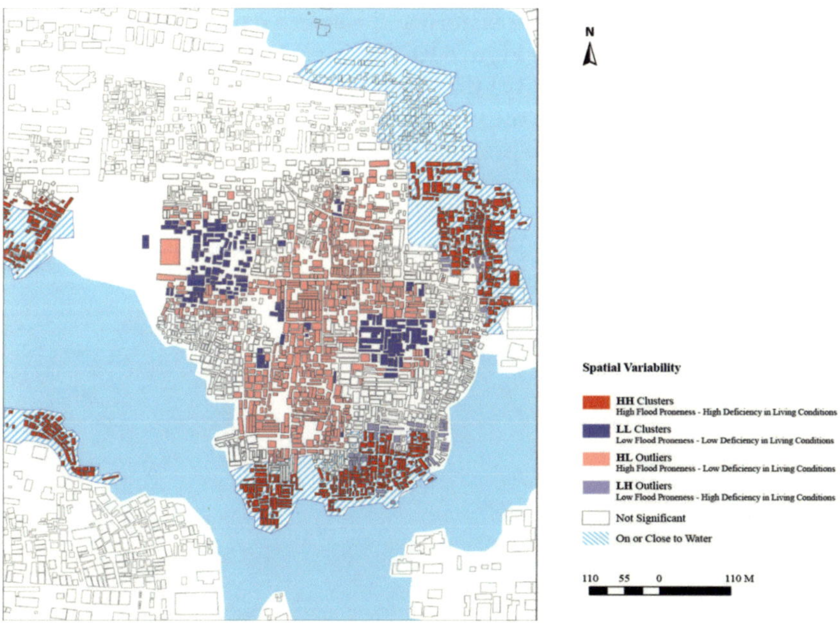

Figure 9.
Spatial variability and clusters in Korail in relation to flood proneness and slum living conditions [by authors].

are the ones presenting the highest risk of flood and the highest deficiency in living conditions—in other words, **HH** clusters.

As a result, flood risks and the proximity to a water body seem to have a considerable influence on the living conditions of the settlements. Concomitantly, the flood risks do not seem to permanently affect the living conditions of a settlement located further away from a water body, as demonstrated by the **HL** outlier. However, this may be due to its central location, in Korail's case.

In effect, this observation can be explained by the fact that social centers are more likely to be located in central and inland locations in the slum than toward the waterfronts to offer a more efficient catching area. For that reason, the presence of such outlier may not be a constant in all slums located close to a water body, and further investigation on this topic would be needed.

In any cases, outlier still experienced a strong reduction in living conditions during floods [15, 35] with the destructions [25] and disruption of the activities [28], including the social centers and the spread of diseases [26].

4.2 Other findings

Besides the examination of the relationship between the Flood Proneness and the Slum Living Conditions Indexes, a temporal spatial pattern analysis was made to investigate the relationship between the proximity from the water body and settlement's density. The purpose of this second analysis is double. Firstly it will serve to observe where the population is concentered in Korail, and secondly the temporal analysis will also help to determine if the slum expands toward Banani Lake as observed by Olthuis et al. [1].

Hence, the significance map in **Figure 10** shows that, in the slum, multiple settlements (in built units prior to 2011) in various locations exhibit significant correlation in terms of their density and distance from water body. Thus, several cluster formations can be observed in built units located close to the lake. Among them two are significant in the southern and northwestern part of the slum, and two smaller ones are located in the eastern and western part.

The spatial variability map in **Figure 10**, for its part, displays several spatial outliers *LH* (low distance-high density) that capture clusters of built unit presenting a high density and located close to the water body. Most of these clusters correspond to the built units identified in the previous figure.

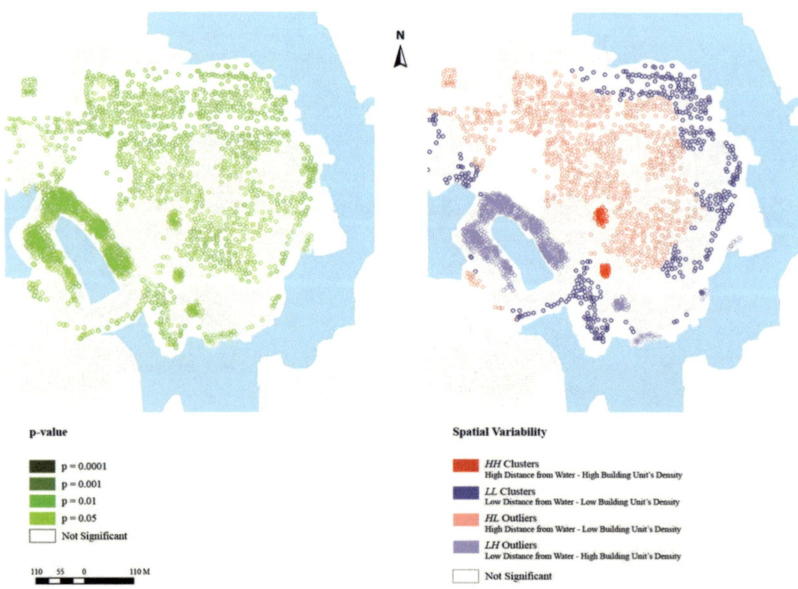

Figure 10.
Significance map (Local Moran's I) for built settlement prior to 2011 (left) – Spatial variability map (Local Moran's I) for built settlement prior to 2011 (right) [by authors].

In addition, this map also illustrates the fact that most of the dense settlements existing in Korail prior to 2011 are located close to the water bodies, surrounding the edges of the lake. Moreover, these areas are also the one presenting a high flood risk and a high deficiency in living conditions in **Figure 9**. As a result, the density of the built units increased as the distance to the water body decreased due to a negative correlation. This phenomenon can be an indicator of the slum's growth toward the water.

As for the dense settlements located further away from the lake, they are mainly localized in areas presenting a high flood risk and a low deficiency in living conditions (**Figure 9**) and are in the buffer zones of the social centers (see Appendix A.2).

The significance map in **Figure 10** showcases that the built unit development between 2011 and 2016 rapidly expanding toward the lake in the western part of the slum. Similar to the observation made prior 2011, this cluster shows significant influence of distance over density of the settlement.

This western expansion can be attributed to the previous observation, in which built units were rapidly expanding in the southern banks, which over a period of time left limited space for further encroachment. Therefore, Korail's expansion leads toward the Banani Lake.

The spatial variability map on **Figure 11** confirms the tendencies observed on the one in **Figure 10**: most of the dense settlements built between 2011 and 2016 in Korail are located close the lake. During this period, the expansion of the slum was mainly focused in the western banks and has steadily consumed the vast majority of area belonging to the lake. Other *LH* (low distance-high density) spatial outliers are also identified in the southern part of the slum around the *LH* outlier observed before 2011. This confirms the hypothesis of a transfer of the slum expansion from the southern part to the western one due to the disappearance of space for further encroachment.

Concerning the *HH* clusters (high distance-high density), the spatial variability map on **Figure 11** also displays similar results as the one in **Figure 10**. These clusters are all concentered in the southern part of the north-south corridor presenting

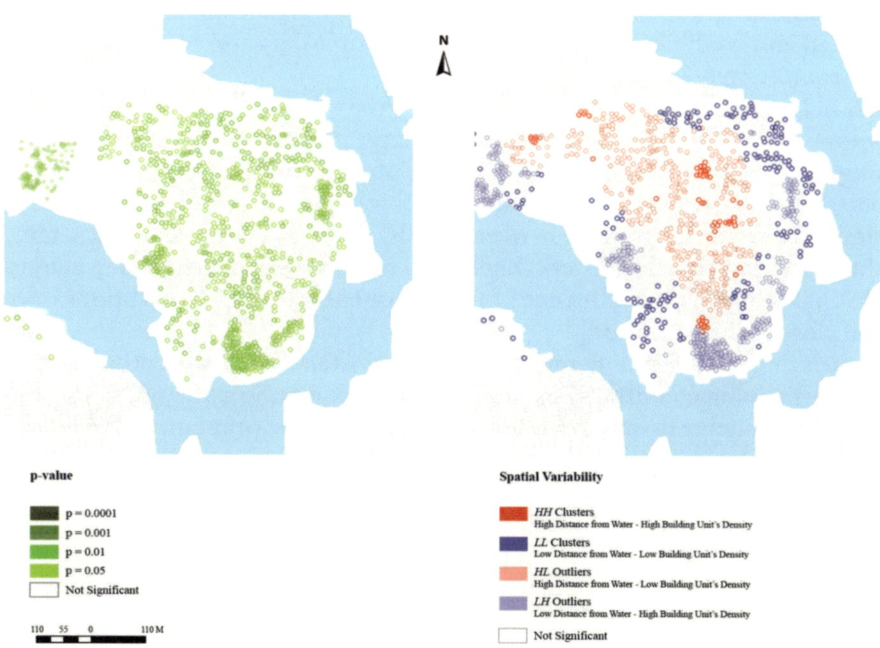

Figure 11.
Significance map (Local Moran's I) for built settlement 2011-2016 (left) – Spatial variability map (Local Moran's I) for built settlement 2011-2016 (right) [by authors].

high risk of flooding and low deficiency in living conditions (**Figure 9**) and are also located in the buffer zones of the social centers (see Appendix A.2).

Consequently, from the observations made from these maps, it appears that most of the dense building units built before 2011 and all the ones built between 2011 and 2016 are located in flood-prone areas and expand toward Banani Lake.

As a result, they are all affected by floods and the resulting hazards of such events. Furthermore, a vast majority of these built units are located close to or on the lake in areas with a high risk of floods and a high deficiency in living conditions. Hence, the living conditions of a large share of the population living in Korail are negatively affected either permanently or regularly by water.

Thereby, Korail can be defined as a **wetslum**—a slum or a large part of a slum where its proximity to a water body and the related flood risks considerably affect the living conditions of its inhabitants. In other words a **wetslum** is a slum where a considerable correlation between flooding and living conditions exists. Thus, **wetslums** are specific human habitats requiring specific environmental and locational management.

5. Conclusions

The framework conceptualized in this paper is inspired by the "living conditions diamond" developed by Gulyani and Bassett [2] as well as the "slum index" introduced by Weeks et al. [4]. They instilled the development of the Slum Living Conditions Index as well as the Flood Proneness Index used to quantify the spatial variability of vulnerability within Korail. Then, clusters were identified through these indexes, and the relationships between the two indexes were examined with a spatial autocorrelation through GeoDa.

It was then used to assess the overall influence of water on the living conditions of a slum. In addition, spatial autocorrelation was also conducted with a temporal analysis to assess the cluster's density with respect to their distance from the water body.

Nevertheless, the study performed in this paper to test the framework necessitated the adjustment of data for the construction of the indexes in reason of the lack of updated and credible data. As a result open-source data and secondary data from several sources were implemented.

The maps produced for the visualization of the two indexes (**Figures 6** and 7) show spatial disparities inside Korail for the exposition to flood risks as well as concerning the living conditions. These observations are confirmed by **Figure 9** where the clusters identified demonstrate that Korail is highly heterogeneous: results that are similar to the findings made in Accra [5, 30]. Indeed, **Figure 9** displays, for building units within **HH** clusters (high flood proneness-high deficiency in living conditions), a significant influence of flood proneness on living conditions in areas located close to the water body.

At the same time, building units within **HL** clusters (high flood proneness-low deficiency in living conditions) exist further away from the water body. These areas have a low deficiency in living conditions in reason of the proximity of several services—social centers in the present case. However, they are still impacted by flooding and thus experiencing periods of rupture in service provision, resulting in the increase of deficiency in living conditions.

As for the **LL** clusters (low flood proneness-low deficiency in living conditions), they are located in more elevated areas, where the rare **LH** clusters (low flood proneness-high deficiency in living conditions) are located in elevated areas close to the water body.

Therefore, these observations confirm the crucial role that water plays on the living conditions of such human habitat as well as the influence of a build unit's localization, such as its distance from a water body and physical factors, such as topography, within a slum. In addition, slums similar to Korail, in terms of flood risk and living conditions, are naturally dynamic—constantly evolving in terms of shape and size [1]. Indeed, **Figures 10** and **11** as well as researches [5, 25] show that Korail's population is growing and that this growth leads to the expansion of the slum toward Banani Lake's bank or directly on it, in the form of dense settlements. Furthermore, the comparison of **Figures 10** and **11** also exhibits the fact that this expansion takes place on any available space, thus explaining the shift observed in the new settlement location from the southern part (prior to 2011) to the western part (2011–2016) of the slum.

As a result, the specific human habitats that are **wetslums**—slums where a considerable correlation between flooding and living conditions exists—are likely to expand toward a water body, thus increasing the number settlements where water affects the living conditions - **HH** clusters.

Hence, this continuous evolution is making it increasingly difficult to provide appropriate interventions and upgrading measures. In that respect, contemporary geo-information and remote sensing technique has enabled researchers to interpret water condition and flood characteristic in an efficient manner. This creates an environment for intercomparison of multinational slums and extent to which their living conditions are affected by water bodies (wetslumness) based on satellite imagery and secondary data. However, such interpretation is due to the lacks of local data that prevent researchers to have a clear on to date vision of the studied settlement.

Thus, the framework conceptualized in this research allows the study of the dynamic characteristic of slums as well as determines locations which require critical upgrading. Indeed, understanding the spatial variability of slums is of crucial importance for slum physical upgrading projects, particularly for **wetslums** where the impact of flooding becomes a key factor which needs to be properly incorporated in order to ensure the success of any physical upgrading project.

5.1 Recommendations

Studying slums at a local scale provides a broader understanding on the organic growth of slums. This level of study captures the physical (topographic) characteristics that go on to influence the living conditions of that settlement. Slum physical upgrading projects need to acknowledge these characteristics in order to achieve sound success and target the right locations for intervention. Place-based upgrading measures will provide relief to slums that require immediate attention.

Slums exhibiting diverse neighborhoods (heterogeneous clusters) require dynamic physical upgrading interventions, which can adapt to the ever-changing environment of these human habitats. For instance, **wetslums** undergo seasonal constraints. They are exposed to heavy floods during certain season, while during the rest of the year, they undergo a dry spell. Dynamic physical upgrading must be able to adapt to these varying environment.

Another obstacle that decision-makers and slum welfare organizations alike are faced with is their ability to record and observe the effect of upgrading measures in real time. It is critical to evaluate slum physical upgrading projects and their positive effect (if any) on the inhabitants, which allows these projects to thrive and/or advance. Hence, there is a growing need to monitor and assess the influence of physical upgrading projects that can be done through modern geo-information platforms such as GPS surveys and web-based platforms for interview of stakeholders.

In short, living conditions in **wetslum** are significantly influenced by their flood proneness, localization, and topographic characteristics. As a result, these particular human habitats require specific physical upgrading projects in terms of dynamic intervention. Also, it is crucial to monitor such dynamic interventions post-implementation, in order to evaluate their real-time effects on the slum inhabitants.

Conflict of interest

None declared.

A. Appendices

A.1 Multi-distance spatial cluster analysis using Ripley's K function

See **Figure 12** and **Table 1**.

The observed K value is larger than the expected K value for the first four classes (distance bands): this proves the settlement pattern is more clustered than random at these distance (scale of analysis). Moreover, the observed K value is larger than the upper confidence envelope value (HiConfEnv), thereby proving the spatial clustering for these distance is statistically significant.

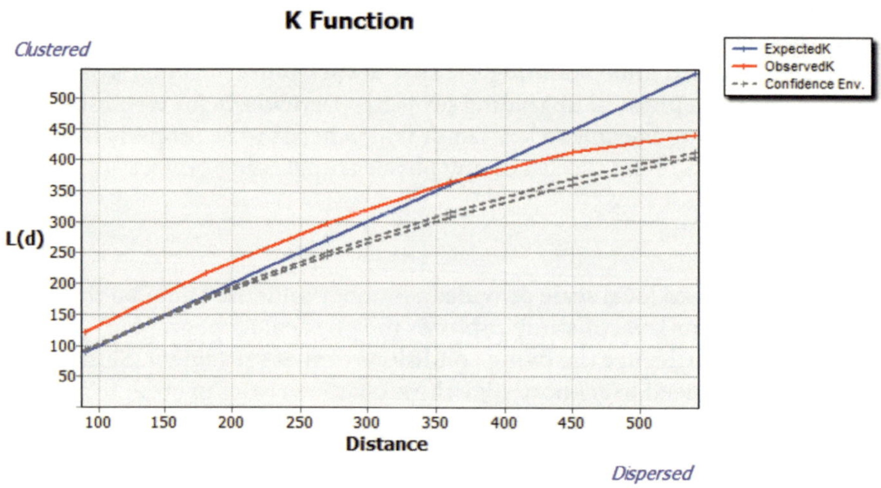

Figure 12.
Multi-distance spatial cluster analysis (Ripley's K function) [by authors].

Class	Expected K	Observed K	Diff K	LwConfEnv	HiConfEnv
0	90.00	121.25	31.25	91.30	92.84
1	180.00	217.21	37.21	172.93	177.08
2	270.00	297.82	27.82	245.12	251.23
3	360.00	364.51	4.51	307.67	315.45
4	450.00	411.80	−38.20	360.93	369.47
5	540.00	441.80	−98.20	404.88	412.90

Table 1.
Multi-distance spatial cluster analysis (Ripley's K function) [by authors].

A.2 Location and buffer zones of the education and health centers in Korail

See **Figure 13**.

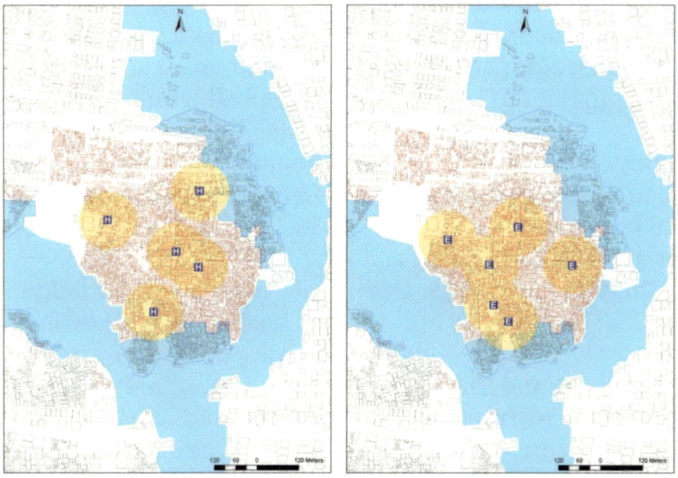

Figure 13.
Health centers and their buffer zone (on the left); education centers and their buffer zone (on the right) [by author].

Author details

Koen Olthuis[1,2]*, Kasirajan Mahalingam[2], Pierre-Baptiste Tartas[2] and Chris Zevenbergen[1]

1 Flood Resilience Group, UNESCO-IHE, Delft, The Netherlands

2 Waterstudio.NL, Rijswijk, The Netherlands

*Address all correspondence to: koen@waterstudio.nl

References

[1] Olthuis K, Benni J, Eichwede K, Zevenbergen C. Slum upgrading: Assessing the importance of location and a plea for a spatial approach. Habitat International. 2015;**50**:270-288. DOI: 10.1016/j.habitatint.2015.08.033

[2] Gulyani S, Bassett EM. The living conditions diamond: An analytical and theoretical framework for understanding slums. Environment and Planning A. 2010;**42**(9):2201-2219. DOI: 10.1068/a42520

[3] Sori ND. Identifying and Classifying Slum Development Stages from Spatial Data [thesis]. Enshede: Faculty of Geo-Information Science and Earth Observation of the University of Twente; 2012

[4] Weeks JR, Hill A, Stow D, Getis A, Fugate D. Can we spot a neighborhood from the air? Defining neighborhood structure in Accra, Ghana. GeoJournal. 2007;**69**(1-2):9-22. DOI: 10.1007/s10708-007-9098-4

[5] Jankowska MM, Weeks JR, Engstrom R. Do the most vulnerable people live in the worst slums? A spatial analysis of Accra, Ghana. Annals of GIS. 2011;**17**(4):221-235. DOI: 10.1080/19475683.2011.625976

[6] Kuffer M, Pfeffer K, Sliuzas R. Slums from space—15 years of slum mapping using remote sensing. Remote Sensing. 2016;**8**(6):1-29. DOI: 10.3390/rs8060455

[7] Hossain S. Social characteristics of a megacity: A case of Dhaka City, Bangladesh. In: Proceedings of the TASA Conference 2006; 4-7 December 2006; University of Western Australia & Murdoch University. Australia: Perth; 2006. pp. 1-9

[8] Adelekan IO. Vulnerability of poor urban coastal communities to flooding in Lagos, Nigeria. Environment and Urbanization. 2010;**22**(2):433-450. DOI: 10.1177/0956247810380141

[9] Douglas I, Alam K, Maghenda M, Mcdonnell Y, Mclean L, Campbell J. Unjust waters: Climate change, flooding and the urban poor in Africa. Environment and Urbanization. 2008;**20**(1):187-205. DOI: 10.1177/0956247808089156

[10] Oduwaye L, Ilechukwu V, Yadua O. Socio-economic determinants of urban poor housing types in Makoko Area, Lagos. In: Proceedings of the REAL CORP 2011 16th International Conference on Urban Planning and Regional Development on the Information Society; 18-20 May 2011; Essen. Schwechat: CORP; 2011. pp. 873-882

[11] Faldetta KF, Reighard DA, Dickinson KL, Wang CQ, George DR, Rodriguez Benavides L, et al. Assessing domestic water quality in Belén municipality, Iquitos, Peru. Journal of Water, Sanitation and Hygiene for Development. 2014;**4**(3):391-399. DOI: 10.2166/washdev.2014.051

[12] Johnson MAS, Smith H, Joseph P, Gilman RH, Bautista CT, Campos KJ, et al. Environmental exposure and leptospirosis, Peru. Emerging Infectious Diseases. 2004;**10**(6):1016-1022. DOI: 10.3201/eid1006.030660

[13] Kumar Karn S, Harada H. Field survey on water supply, sanitation and associated health impacts in urban poor communities—A case from Mumbai City, India. Water Science and Technology: A Journal of the International Association on Water Pollution Research. 2002;**46**(11-12): 269-275. DOI: 10.2166/wst.2002.0749

[14] Heryana A, Setyawan D, Setiawan B, Purnama DH. Flood hazard analysis as impact of climate change on slum areas in

Palembang, South Sumatera. Journal of Sustainable Development. 2015;**8**(1):218-225. DOI: 10.5539/jsd.v8n1p218

[15] Jabeen A, Johnson C, Allen A. Built-in resilience learning from grassroots coping strategies for climate variability. Environment & Urbanization. 2010;**22**(2):415-431. DOI: 10.1177/0956247810379937

[16] Simon RF, Adegoke AK, Adewale BA. Slum settlements regeneration in Lagos Mega-city: An overview of a waterfront Makoko community. International Journal of Education and Research. 2013;**1**(3):1-16. Available from: https://www.ijern.com/images/March-2013/16.pdf [Accessed 2016-06-02]

[17] Oskam V. Sanitation Situation in Central Davao City Slum Areas. 2012. Available from: https://issuu.com/vincentoskam/docs/2012-12-21_-_c.02_-_sporadic_indicative_sanitation [Accessed 2016-06-02]

[18] Abiko A, de Azevedo Cadoso LR, Rinaldelli R, Riogi Haga HC. Basic costs of slum upgrading in Brazil. Global Urban Development Magazine. 2007;**3**(1):1-23. Available from: https://www.globalurban.org/GUDMag07Vol3Iss1/Abiko%20PDF.pdf [Accessed 2016-06-02]

[19] Flores-Fernandez R. Physical and spatial characteristics of slum territories vulnerable to natural disasters|. Les Cahiers de l'Afrique de l'Est. 2011;**44**:5-22. Available from: https://halshs.archives-ouvertes.fr/halshs-00751670/document [Accessed 2016-06-02]

[20] Scovronick N, Lloyd S, Kovats S. Climate and health in informal urban settlements. Environment & Urbanization. 2015;**27**(2):657-678. DOI: 10.1177/0956247815596502

[21] UNISDR. Summary and Recommendations: 2009 Global Assessment Report on Disaster Risk Reduction; Risk and Poverty in a Changing Climate. Geneva: United Nations; 2009. 207 p. ISBN/ISSN: 9789211320282

[22] Ballesteros MM. Linking Poverty and the Environment: Evidence from Slums in Philippine Cities. Philippine Institute for Development Studies, Discussion Paper Series N°. 2010-33 [Internet]. 2010. Available from: https://dirp4.pids.gov.ph/ris/dps/pidsdps1033.pdf [Accessed 2016-06-02]

[23] Braun B, Aßheuer T. Floods in megacity environments: Vulnerability and coping strategies of slum dwellers in Dhaka/Bangladesh. Natural Hazards. 2011;**58**:771-787. DOI: 10.1007/s11069-011-9752-5

[24] Altea GJ. Urban Acupuncture & Dwelling Formability, Regeneration of Coastal Informal Settlement in Manila [thesis]. St. Louis: Washington University; 2013

[25] Porio E. Vulnerability, adaptation, and resilience to floods and climate change-related risks among marginal, riverine communities in Metro Manila. Asian Journal of Social Science. 2011;**39**(4):425-445. DOI: 10.1163/156853111X597260

[26] Aman J, Rahman NN, Zahir S. "Beltola lilies"—A solution of housing for lower income people and introduction of a module for flood areas. Creative Space. 2016;**3**(2):119-131. DOI: 10.15415/cs.2016.32008

[27] Rashid H, Hunt LM, Haider W. Urban flood problems in Dhaka, Bangladesh: Slum resident's choices for relocation to flood-free areas. Environmental Management. 2007;**40**:95-104. DOI: 10.1007/s00267-006-0233-7

[28] Kanke Arachchilage KJ. Reducing the Vulnerability of the Urban Poor

to Climatic Change: Experiences from Colombo and Dhaka [thesis]. Vancouver: Faculty of Graduate and Postdoctoral Studies (Architecture), University of British Colombia; 2013

[29] Rao BT, Thakur JS. Vulnerability assessment in slums of union territory, Chandigarh. Indian Journal of Community Medicine. 2007;**32**(3):189-191. DOI: 10.4103/0970-0218.36824

[30] Jankowska M. Continuums of Vulnerability in the Slums of Accra, Ghana. The World Bank [Internet]. 2011. Available from: http://siteresources.worldbank.org/INTURBANDEVELOPMENT/Resources/336387-1272506514747/Jankowska.pdf [Accessed 2016-06-02]

[31] Melecky L. Spatial autocorrelation method for local analysis of The EU. Procedia Economics and Finance. 2015;**23**:1102-1109. DOI: 10.1016/S2212-5671(15)00484-0

[32] Fan C, Myint S. A comparison of spatial autocorrelation indices and landscape metrics in measuring urban landscape fragmentation. Landscape and Urban Planning. 2014;**121**:117-128. DOI: 10.1016/j.landurbplan.2013.10.002

[33] Hammadi S. The Ignored Population. New Age: The Outspoken Daily [Internet]. 2014. Available from: http://newagebd.net/29880/the-ignored-population/ [Accessed 2016-06-02]

[34] Iftekhar Ahmed K. Urban Poor Housing in Bangladesh and Potential Role of ACHR. Asian Coalition for Housing Rights (ACHR), Bangkok, Thailand [Internet]. May 2007. Available from: http://www.achr.net/upload/downloads/file_22122013020636.pdf [Acessed 2016-06-02]

[35] Rabbani G. Climate Change Vulnerabilities for Urban Areas in Bangladesh: Dhaka as a Case. Resilient Cities [Internet]. 28 May 2010. Available from: http://resilient-cities.iclei.org/fileadmin/sites/resilient-cities/files/docs/B4-Bonn2010-Rabbani.pdf [Accessed 2016-06-02]

[36] Cameron S. Access to and Exclusion from Primary Education in Slums of Dhaka, Bangladesh. CREATE Pathways to Access, Research Monograph No. 45. Flamer: University of Sussex, Centre for International Education; 2010. p. 42 ISBN: 0-901881-52-X

[37] Degert I, Parikh P, Kabir R. Sustainability assessment of a slum upgrading intervention in Bangladesh. Cities. 2016;**56**:63-73. DOI: 10.1016/j.cities.2016.03.002

[38] Asian Development Bank. Urban Primary Health Care Services Delivery Project [Internet]. 2015. Available from: https://www.adb.org/projects/42177-013/main [Accessed 2016-06-02]

[39] Stevenson R, Sarker A, May MA. Frugal map making: Experiences from Korail. BRAC [Internet]. 2013. Available from: http://blog.brac.net/2013/09/frugal-map-making-experiences-from-korail/ [Accessed 2016-06-02]

[40] Bangira T. Mapping of Flash Flood Potential Areas in the Western Cape (South Africa) Using Remote Sensing and in situ Data [thesis]. Enshede: Faculty of Geo-Information Science and Earth Observation, University of Twente; 2013

[41] Bates PD, De Roo APJ. A simple raster-based model for flood inundation simulation. Journal of Hydrology. 2000;**236**(1-2):54-77. DOI: 10.1016/S0022-1694(00)00278-X

[42] Tarboton DG, Sazib N, Dash P. Quick start guide to using the TauDEM ArcGIS toolbox. Utah State University [Internet]. 2015.

Available from: http://hydrology.usu.edu/taudem/taudem5/TauDEM51GettingStartedGuide.pdf [Accessed 2016-06-02]

[43] Tarboton DG. A new method for the determination of flow directions and contributing areas in grid digital elevation models. Water Resources Research. 1997;**33**(2):309-319. DOI: 10.1029/96WR03137

[44] Anselin L. Exploring Spatial Data with GeoDaTM: A Workbook. Chicago: Center for Spatially Integrated Social Science [Internet]. 2005. Available from: https://www.csiss.org/clearinghouse/GeoDa/geodaworkbook.pdf [Accessed 2016-06-02]

[45] Zhang C, Luo L, Xu W, Ledwith V. Use of local Moran's I and GIS to identify pollution hotspots of Pb in urban soils of Galway, Ireland. Science of the Total Environment. 2008;**398**:212-221. DOI: 10.1016/j.scitotenv.2008.03.011

Chapter 7

Wet and Dry Spells over Southeast Peninsular India

Mohana S. Thota

Abstract

The southeast peninsular India contains, to name a few, several important cities crucial for trade and economic growth of the country, rice bowls, institutes for science and technology, space port, etc. Despite its importance, not many reports exist on rainfall and its variation on different temporal scales over this region during southwest monsoon, partly because the rainfall in this region is relatively less and it forms only a minor part of all India rainfall. Here, an attempt has been made to understand differences in thermal and dynamical characteristics and energetics of the atmosphere between wet and dry spells of the Indian summer monsoon over the southeast India by utilizing various observations and reanalysis products. Observations demonstrate that the difference in the thermal structure between wet and dry spells is significant only in the lower troposphere (< 2-3 km) with mean CAPE values are reaching as much as 1000 Jkg-1during wet spell. Vertical buoyancy profiles indicate the bi-modal distribution during dry spells with peaks in 700 and 500 hPa levels. The observed thermal features are not confined to Gadanki but seen over entire southeast peninsular India. Associated dynamical variations also exhibit obvious differences during wet and dry spell. The diurnal variation of winds exhibits difference in amplitude and phase are remarkably large during dry spell than in wet spell. Synthesis are all the measurements indicates that the thermal and dynamical differences observed in wet and dry spells are pronounced in the boundary layer.

Keywords: wet and dry spells, Monsoon, CAPE, Southeast peninsular India

1. Introduction

The term "monsoon" is an Arabic word, which means seasonal reversal of wind direction. Prevailing wind direction between winter and summer seasons are the basis for delineation of the regional monsoon around the world. According to [1], the monsoonal region in the tropics extends between 25S–35N and 30 W–170E. This broad region comprises of three parts, the African monsoon and South and East Asian monsoon systems. From past several decades, these monsoon systems acquired fervent attention by scientific community and common people as well. These are globally distributed atmospheric phenomena giving surplus amount of water to mankind. It is estimated that the monsoon rains all over the world and provides ~60% of global water supply [2]. Among the monsoon systems over the globe, the Asian monsoon is the largest, and Asian regions are critically influenced by the evolution and inherent variability. It interacts with El Nino/Southern Oscillation and extratropical weather systems and thereby controls the global circulation through teleconnections. Further, monsoon variability is also influenced by aerosols, tropical typhoon activity, etc. The Asian monsoon can be classified into two types: (1) South Asian monsoon or Indian monsoon, which affects the Indian subcontinent and surrounding regions and (2) East Asian monsoon which affects the countries like China, Korea, and Japan. In the present chapter, we primarily focus on the Indian monsoon especially over southeast India.

2. Indian monsoon: Southeast peninsular India

Indian subcontinent is situated at the vicinity of the monsoonal region defined by [1]. Copious amount of rainfall occurs over the Indian sector during June–September, also known as summer (or southwest) monsoon season. It accounts for ~75–80% of the annual rainfall over major parts of the Indian subcontinent. Agrarian countries like India heavily depend on this rainfall. Any deficit in the seasonal rainfall will have an adverse effect on the agriculture and economy of the country. The commencement of rainy season over Indian subcontinent is distinguished by the widespread rainfall over Kerala coast in late May/early June. In general monsoon sets over Kerala on June 1 with a standard deviation of 8 days [3]. Monsoon arrival to the central India takes 10–15 days after the onset and completely occupies the subcontinent by mid-July [15]. The withdrawal of monsoon takes place in September from northwest India, and by October 20, retreat of the southwest monsoon completes, and another monsoon, called northeast monsoon, sets in over southeast India.

Mean seasonal summer monsoon rainfall is not homogeneous (see **Figure 1**); it varies spatially (within India) and temporally (within the season) over the Indian region. From **Figure 1**, a large spatial asymmetry in seasonal mean rainfall can be seen over the Indian subcontinent. The rainfall occurrence is high over the West coast and Northeastern India (approx. >16 cm), moderate to heavy over central India (8–15 cm), and low over northwestern and southeastern peninsular India (~4 cm). Southeast peninsular India receives considerable amount of rainfall during the northeast monsoon season, i.e., from October

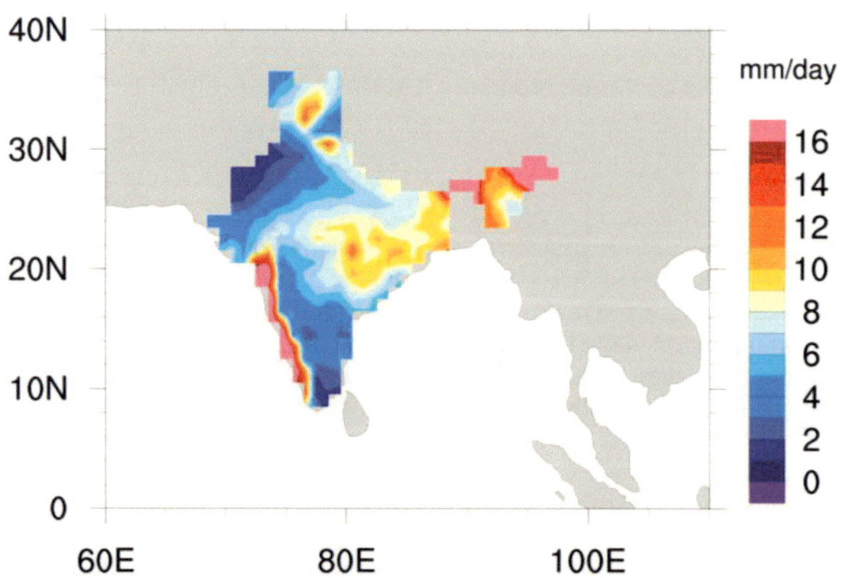

Figure 1.
Spatial distribution of mean seasonal summer monsoon rainfall. The high-resolution 1 × 1 degree rainfall data generated by IMD are used for this plot.

through December. After the onset of the southwest monsoon over Kerala coast, monsoon rainfall and the intertropical convergence zone (ITCZ) move northward. However, the northward propagation or advance of the monsoon is not always smooth; rather it takes place in pulses or epochs, in association with the convective activity [4].

During the southwest monsoon season, the large-scale rainfall distribution is mostly concentrated over two regimes: North Bay of Bengal and south of the equator in Indian Ocean. Winds converge in these regions, and these regions become most favored zones for convection and the formation of the tropical convergence zone (TCZ). Mean position of the TCZ is not constant; instead it oscillates with different periodicities and scales. Oscillations of the TCZ are synchronized with enhanced and suppressed convective activity known as active and break phases of the monsoon. Thus, the monsoon is a manifestation of the seasonal migration of the ITCZ, and monsoon variability is associated with the space–time variation of the ITCZ [5]. As discussed earlier, the seasonal monsoon rainfall varies spatially and temporally, and the variation of rainfall depends mainly on the duration and the time of occurrence of active and break phases. Natural disasters caused by the extreme hydrometeorological events are manifestation of intraseasonal variability (ISV) [6–8].

3. Intraseasonal variability

The day-to-day variability in the rainfall plays an important role in deciding the seasonal rainfall. Seasonal mean monsoon rainfall is affected by the occurrence and strength of the active and break spells [6]. It is well known from the earlier studies that the modes of 30–60 days and 10–20 days are most important in controlling the Indian summer monsoon rainfall (ISMR) [5, 9–13]. [14] studied the spatial variability of intraseasonal oscillations (ISOs) in deficit and excess monsoon years and demonstrated the dominance of 30–60 days over west coast and southeast region during the deficit monsoon year, while excess monsoon years are characterized by high-frequency synoptic (3–5 days) oscillations. Another important result obtained by them is the weakening of 30–60-day oscillation and strengthening of 10–20-day oscillation over central India and some parts of the west coast.

Active and break spells of Indian monsoon associated with intraseasonal oscillations (ISOs) need to be understood properly as they control the seasonal rainfall. It is noted by several researchers that excess monsoon rainfall years generally have more active spells and deficit monsoon rainfall years will have prolonged and/or more break spells (see [15–17]). A modest decrease in the monsoon rainfall (e.g., 10% of the long-term mean) significantly affects the food productivity (see [15, 18–21]). The knowledge of active and break spells on a regional scale is more crucial and important than all-India integrated active/break spells for agriculture and water management sectors. Therefore, the prediction of occurrence of active/break spell and their duration is much more important for the management of sowing and harvesting than the seasonal mean rainfall. In the recent past, several research works have been carried out to understand and identify the active and break spells over the Indian subcontinent using various parameters (see [5, 17, 22–28]. Since the definitions and techniques used by the aforementioned authors for the identification of active and break spells are different, the duration and occurrence of the spells in any monsoon season can be different [29]. For instance, [30] reported that there is hardly any overlap between the spells observed by them and those observed by [15]. Ramamurthy [23] identified breaks over monsoon zone using the rainfall data from 1888 to 1967 and inferred that according to his criteria, there were no break days observed in 10 consecutive years. However, the study reported that the duration of break days varies from 3 to 15 days with 30% of the spells longer than or equal to 7 days.

4. Spatial variability of active and break spells

Earlier studies have shown that intraseasonal variations in rainfall are not coherent over the Indian region and the active/break spells of some subdivisions are in opposite phase with each other [16, 31, 32]. For instance, the spatial structures of the active/break spells are organized in such a way that southeast peninsular and northeastern parts of India exhibit an out-of-phase relation with the monsoon zone or central India. Also, rainfall increases near the foothills of the Himalayas, following the northward movement of monsoon trough, during break spells for the monsoon zone. Therefore, knowledge of active and break spells on a regional scale is important as different regions grow different crops and follow different practices.

In a study, [33] demonstrated that there exists large spatial variability in active and break spells using 58 years (1951–2007) of high-resolution (1° × 1°) rainfall data generated by the India Meteorological Department (IMD) [16]. They generated spatial maps for rainfall fraction (in terms of %) corresponding to break periods as defined by different criteria, discussed above [17, 25–27] (**Figure 2**). **Figure 3** shows that large rainfall fraction of ~20–40% occurs over southeast India during all-India break. The clear discrepancy in the rainfall distribution between these regions indicates the need for identifying active and break spells separately for the southeastern region of India.

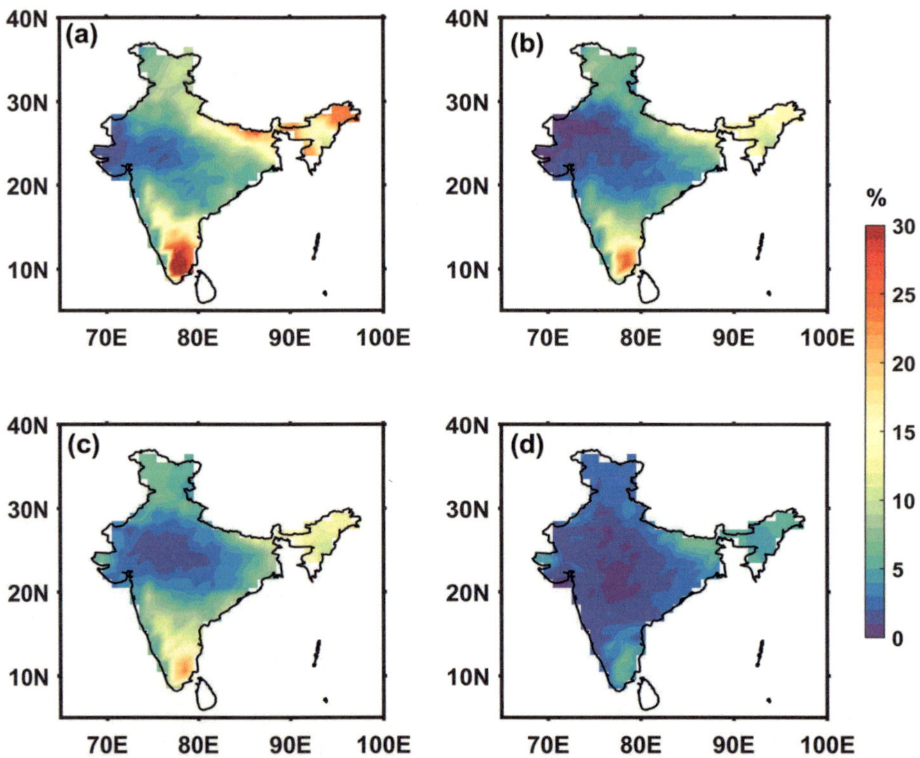

Figure 2.
Spatial distribution of rainfall fraction (%) in all-India break periods, as defined by (a) [25], (b) [27], (c) [17], and (d) [26]. Reproduced from ***Figure 1****, [33].*

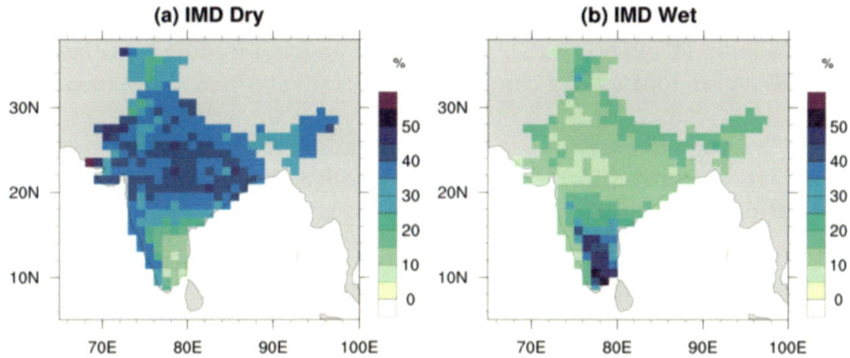

Figure 3.
Rainfall percentage contribution during (a) dry and (b) wet spells of the southwest monsoon from 1951 to 2015 using IMD 1 × 1 deg. gridded data.

4.1 Identification of wet and dry spells

In the present chapter, the terms "wet" and "dry" spells are used instead of active and break spells, in order to avoid confusion with all-India active and break spells of the summer monsoon rainfall. The method of analysis for the identification of wet and dry spells is discussed below. Wet and dry spells are identified using area-integrated surface rainfall measurements, following [16]. The area over which the surface rainfall is integrated (9.5°–15.5° N and 77.5°–81.5° E) is selected based on the correlation analysis. This is obtained by considering the correlation between time series of rainfall at each grid point and areal averaged rainfall. Only those grids with correlation coefficient >0.5 are considered. Following the procedure described in the above section, a total of 943 from 72 dry spells and 391 from 54 wet spells are identified from the 15 years (1995–2009). The spells are ranging from 5 to 50 days during dry and from 3 to 31 days during wet. Average time span during wet spell is ~7, whereas the span is ~13 days during the dry spell. Among dry (wet) spells, ~34 (48) % of spells have time span longer than the average length of dry (wet) spell.

It is general belief that the rainfall in the rain-shadow region of southeastern peninsular India occurs in isolated convective storms or along the coast mainly due to sea-breeze intrusions. To examine how much rainfall is due to large-scale systems (in wet spell) and how much is due to isolated storms (in dry spell), the rain amount contribution by each spell to the seasonal rainfall is estimated (**Figure 3**).

As mentioned above, the wet and dry spells are in opposite phase in the monsoon zone and southeast peninsular India, i.e., during wet spell, southeastern peninsular India gets good amount of rainfall; nevertheless the monsoon zone seldom gets rainfall and vice versa. Though the wet spell persists only 22% of time in the southwest monsoon, 50–60% of the seasonal rainfall occurs in that spell. On the other hand, only 5–20% of seasonal rainfall occurs in the dry spell. One can see clearly that good amount of rainfall (**Figure 3b**) occurs along the west coast in dry spell and almost no rain in the eastern side of the Western Ghats in the southern peninsula (south of 15° N) in wet spell.

5. Salient features of wet and dry spells

To know the basic differences between these wet and dry spells, it is necessary to understand the thermal and dynamical characteristics associated with these spells. To better understand these processes, observational data collected with a suite of unique instruments at the National Atmospheric Research Laboratory (NARL), Gadanki (13.45 N, 79.2E), located over Southeast India is utilized. It is a semiarid region located in a complex hilly terrain at 375 m above mean sea level (MSL).

In the following subsections, a brief description of the basic characteristics of wet and dry spells is given. For those who are interested can find enough evidence and description in the research papers referenced therein.

5.1 Thermal characteristics

A comprehensive analysis is performed onto the radiosonde data launched from Gadanki twice daily during synoptic hours (00 GMT and 12 GMT) along with surface automatic weather station (AWS) data which has revealed several interesting facts during these spells. Majority of the soundings (~96%) reached above tropopause level and encouraged us to study the vertical variations of temperature (T) and moisture (q).

Figure 4 shows the variations of surface T and q between wet and dry spells are in the form of histograms (in terms of % occurrence). It is apparent from the figure that the wet spells are relatively moist and cooler than dry spells. Although the surface T and q distribution in these spells show some overlapping, the mean (mode) values differ by 1.6 K (~2 K) in T and ~2 g kg^{-1} in q. Note that the variations in T and q are not biased by the seasonal variations in T and q over the study region; rather, their difference in spells are mainly caused by the active/subdued convection in the monsoon season.

Next, to understand how convection in wet spells changes the vertical structure of thermodynamic variables through complex feedbacks (latent heating, etc.), the difference between wet and dry spells of composites of T ($T = Td - Tw$), q ($q = qd - qw$), and equivalent potential temperature, θe ($\Delta\theta e = \theta ed - \theta ew$), is given in **Figure** 5. Suffixes "d" and "w" denote dry and wet spells, respectively.

Maximum temperature differences are seen in the lower and upper troposphere and are of the order of 1–2 K. The lower troposphere (below ~2 km) and upper troposphere (11–16 km) are warmer in dry spell than in wet spell. Between

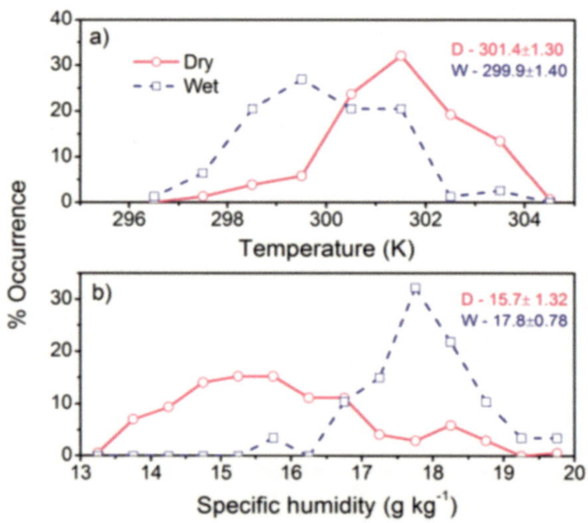

Figure 4.
Frequency distributions (in terms of % occurrence) for the surface (a) temperature and (b) humidity during wet and dry spells. The means of the distribution are also shown in the figure (source from [47]).

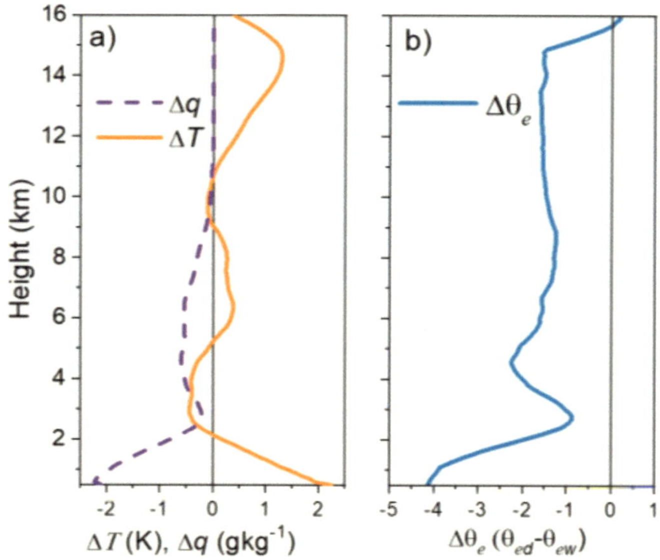

Figure 5.
Vertical profiles of (a) mean temperature, humidity and (b) equivalent potential temperature differences between wet and dry spells ($T = T_d - T_w$, $q = q_d - q_w$, and $\Delta\theta_e = \theta_{ed} - \theta_{ew}$). Suffixes 'd' and 'w' denote dry and wet spells, respectively (reproduced from [47]).

2 and 11 km, the temperature difference between spells is small (the temperature is higher in wet spell than in dry spell in two height regions, 2–5 and 9–10.7 km). The present observations do not show a clear heating in the middle troposphere during the wet spell, in contrast with the earlier studies during the wet spell of the Australian and Indian monsoon systems (over oceans) [34–36]. Similarly, the difference in q composites between wet and dry spells is relatively large below 7 km (>0.5 g kg^{-1}), except in the height region from 2 to 4 km, and small q in this height region is perhaps due to the enhanced detrainment and humidity in dry spell.

Except for this narrow layer, the results are consistent with those obtained in Australia and India, where a moist environment was observed over greater depths during the active phase of monsoon [34–36]. Yet, the role of advection in drying the troposphere during dry spells is not clear. Evaporation of surface moisture can explain the observed humidity differences between spells. Though there are no large local water bodies nearby, evaporation of surface moisture (in general is more during wet spell because of more rain) can enhance humidity in the lower troposphere, as observed in the figure, during wet spell. The other possible candidate for the dryness of the atmosphere during dry spells is large-scale subsidence from higher altitudes.

Next, several potential instability indices (stability index, CAPE, etc.) have been developed to measure the susceptibility of a given temperature and moisture profile to the occurrence of deep convection. Perhaps the most popular and widely used parcel instability parameter is CAPE. The variabilities of the different parcel instability parameters (CAPE, CINE, and stable layers) from wet to dry spells are primarily discussed. **Figure 6a–d** shows the frequency distributions of lifting condensation level (LCL), equilibrium level (EL), CAPE, and CIN during wet and dry spells. Quantitatively, about 71% (only 25%) of the LCL population is larger than 800 hPa during wet (dry) spell. The mean values of LCL for wet and dry spells are, respectively, 827 and 772 hPa. In contrast, EL distributions during wet and dry spells exhibit an opposite trend with cloud systems reaching higher altitudes in wet spells. Among the parcels which reached the EL, about 42 (11) %

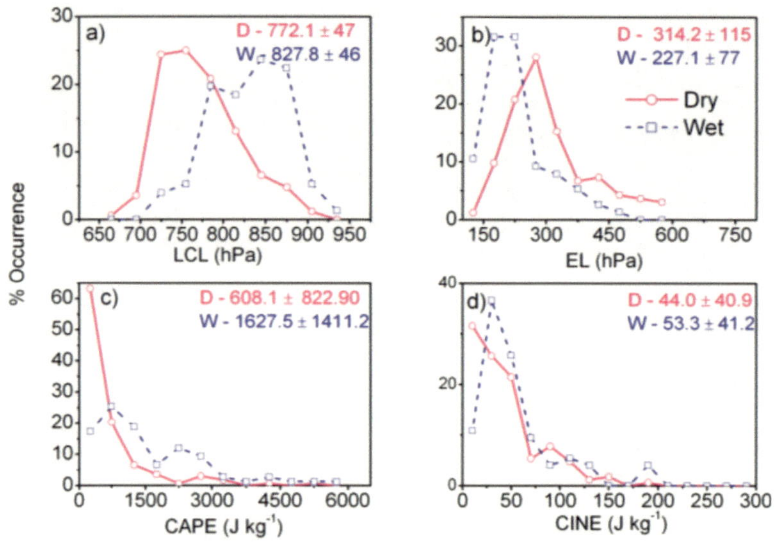

Figure 6.
Frequency distributions (in terms of % occurrence) for (a) LCL, (b) EL, (c) CAPE, and (d) CINE. The means of the distribution are also shown in the figure (source [47]).

of the EL population during wet (dry) spells shows values smaller than 200 hPa with mean altitudes of EL for wet (dry) spells being 227 (314) hPa. Distributions of LCL and EL altitudes indicate that clouds are deeper in wet spell than in dry spell. CAPE distribution and mean CAPE values are different in different spells with large values of CAPE which are seen more frequently in wet spell (**Figure 6**). For example, ~32 (6) % of the CAPE population in wet (dry) spells shows values larger than 2000 J kg^{-1}. Mean CAPE value for wet spells (1627 ± 1411 J kg^{-1}) is larger than for dry spells (608 ± 823 J kg^{-1}), and interestingly standard deviation of CAPE is also large in both spells. Note that the CAPE is estimated from the sounding data available at a fixed time, i.e., ~17:00 LT. Some of the soundings, therefore, represent conditions before active convection and some after convection and others at the time of convection. This perhaps is the main reason why we see significant variation in CAPE values within the spell. The distributions for CINE look similar in both spells (**Figure 6d**) with mean CINE nearly equal (53.3 ± 41.2 and 44 ± 41 J kg^{-1} for wet and dry spells, respectively). Like in the case of CAPE, in both spells, the variability within the spell is comparable to the mean value of CINE for that spell. The range of CINE observed at Gadanki is comparable to that observed over other tropical continental stations (Singapore 1.3 N, 103.8E, [37]) but smaller than that observed over oceans [38]. Although the mean values of CINE are small, they can be considerable on individual days. Note that the present observations were taken during the dusk hours, and therefore the second possibility can be ruled out. If we do not consider any forced lifting, an updraft of ~9 m s^{-1} is required to overcome CINE of ~40 J kg^{-1}. Existence of such strong vertical velocities in non-convective periods is rare. Therefore, in both spells, some external forcing is required to overcome this inhibition energy and to trigger convection.

Contrasting with other studies, CAPE values estimated over tropical oceans (Bay of Bengal and western Pacific) are in contrast to the present observations, with large values in break phase (before convection) and small in active phase [34, 38–40]. The regions influenced by maritime and continental flows also show similar features with larger values of CAPE in mesoscale convective systems associated with the break phase than those with active phase, for example, over Darwin in northern Australia [41] and over Brazil [42]. The main physical reason for the lower values of CAPE in active spells is associated with the decrease of equivalent potential temperature (θe) due to the overturning of the atmosphere during convection [34]. This could be, during wet spell, convective downdrafts induced by precipitation loading, melting, and evaporation bring the relatively dry mid-tropospheric air into the boundary layer and reduces θe near the surface. In addition, the latent heat release in deep clouds warms the middle and upper troposphere and reduces the instability of the atmosphere. All these factors reduce the CAPE during active phase of the monsoon or during the convectively active period. However, at Gadanki, variation in thermodynamical parameters from wet to dry spells is greater near the surface and in the boundary layer. Above the boundary layer and in the middle troposphere (~2.5–11.5 km), the temperature difference between the composites for wet and dry spells is within 0.5 K. Also, in contrast to the expected, larger θe values are observed during wet spell in the lower troposphere [43].

Next, to examine the impact of stable layers in governing the convection growth, **Figure 7** shows the vertical distribution of stable layers in terms of lapse rates (top panel) and percentage occurrence (bottom panel). Following [47], the vertical distribution of % occurrence of temperature lapse rates exceeding certain thresholds (3, 4, and 5 K km^{-1}, all are smaller than the moist atmospheric lapse rate in the troposphere, ~6 K km^{-1}) during dry and wet spells are shown in **Figure 7a** and **b**. Following the procedure described by [47], the % occurrence of stable layers at each altitude is estimated from the ratio between the number of occurrences when the lapse rate exceeds a threshold and the total number of lapse rate data points at that altitude (see [47] for more details). From the analysis it was observed that statically stable layers are predominantly seen in the lower and middle troposphere (below 8 km). The % occurrence distribution for 5 K km^{-1} temperature lapse rate shows a

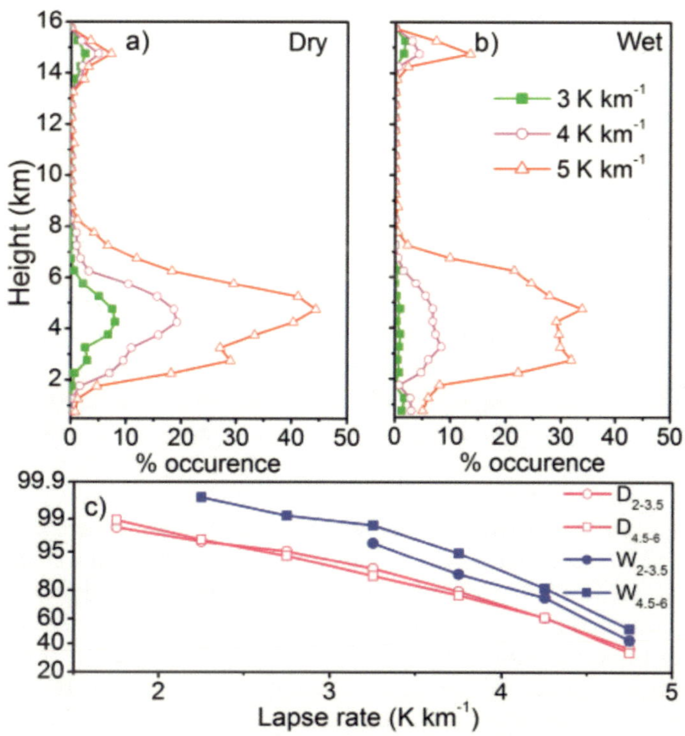

Figure 7.
Vertical distribution of % occurrence of stable layers defined by different temperature lapse rates during (a) dry and (b) wet spells. (c) Frequency distribution of temperature lapse rates during wet and dry spells in two height regions (2–3.5 and 4.5–6 km). D and W denote dry and wet spells, respectively (figure from [47]).

broad distribution between 2 and 7 km with two small peaks centered on 2–3 km and 5 km, while the distributions for other two temperature gradients show a single peak centered around 4.5 km in dry spell and slightly at a lower altitude in wet spell. Nevertheless, the magnitude of % occurrence is different in both spells. The peak in the height region of 2–3 km corresponds very well with the altitude of the boundary layer height and the second peak with the 0°C isotherm level. **Figure 7b** shows the cumulative distribution of the magnitude of stable layers in the height region of 2–3.5 and 4.5–6 km (the height regions in which the % occurrence of stable layers is relatively more) during dry and wet spells. Interestingly, both height regions exhibit strongest stable layers which exist in dry spell. Contrasting the lapse rate distributions in these two height regions indicates that gradients near the freezing level are stronger than their counterparts at low levels. This is true in both spells of the monsoon.

From **Figure 7**, strong inversion layers exist below 6 km height region during dry spell. These stable layers prohibit the growth of convection and modify the distribution of moisture. In other words, the stable layers will enhance the detrainment and thereby moisture near that level; Ref. [33] also supports this view based on draft core statistics. They observed that the shallow cores are more prevalent in dry spell with the maximum percentage occurrence of core tops in the height region of 3–5 km.

5.2 Energetics

In the earlier section, thermal characteristics of wet and dry spells are discussed at a tropical station, situated over southeastern peninsular India. Now, a logical question is raised, whether the observed feature of wet and dry spells is only confined to a station or extended throughout southeastern peninsular India. This issue is discussed in the following section.

For the estimation of CAPE and related thermal parameters, along with the observations, three global reanalysis products are utilized (ERA-interim; MERRA; and NCEP). For more details and description of the data sets (see [44–46]). Though all the gridded data sets were available at 00, 06, 12, and 18 UT, data at 12 UT were only employed, as it coincides with the balloon launch time over the stations.

Since variations in CAPE largely depend on parcel level of origin and its temperature and moisture content as discussed in above section, here, the exercise is repeated to compare with reanalysis products. **Figure 8** shows the composite percentage occurrence of moisture (q) and temperature (T) during wet and dry spells. Wet spells (dashed curves) are relatively moist and cooler than dry spells over Gadanki region, corroborating with [47]. Although the distribution of surface T and q in these spells exhibit reasonably good agreement among the reanalysis products, mean magnitudes differed by 2–3 K in T and ~2 g kg^{-1} in q. Further, a considerable difference between the percentage occurrence of sonde and reanalysis data products is noteworthy. From **Figure 8**, it is observed that surface parcels are more moist and cooler in the reanalysis products than observations, indicating the surface instability in reanalysis are larger than sonde observations. These large differences in the surface parcel thermal properties thus result in large differences in CAPE.

Next, atmospheric stability during wet and dry spells by means of q and T profiles over Gadanki is discussed. Vertical profile of q' over Gadanki from GPS radiosonde observations and reanalysis data are shown in **Figure 9a** and **b**. As expected, it is observed that wet spell q' composites are moist, with magnitudes reaching 1.5 g/kg, compared to dry spell. During dry spell, dryness in the atmosphere, as evidenced by negative q', is extending throughout the tropospheric column below 100 hPa. However, more prominent drying is seen at two levels, one below 800 hPa level and

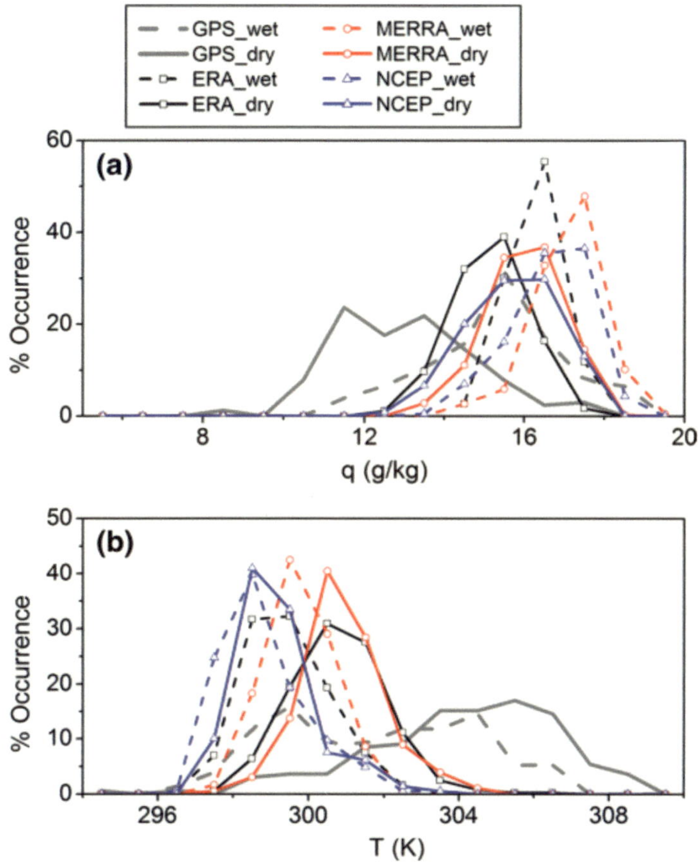

Figure 8.
Histogram of parcel moisture (a) and temperature (b) from GPS sonde observations and reanalysis products during dry (solid line) and wet (dashed line). Different colors in the legend represent the different products (source: [48]).

the other at upper tropospheric levels around 400–500 hPa. Although reanalysis data exhibit a good agreement with the observations, magnitudes of q anomalies slightly vary. For instance, upper tropospheric drying and moistening are not clearly visible in composite profiles of NCEP in both dry and wet spells. From **Figure 9**, it is inferred that changes in the T' are much smaller especially around mid-troposphere levels, with the magnitudes ranging from ±0.5 K in both the spells. However, large changes are seen only at boundary layer heights (below 800 hPa) spell composites (see **Figure 9c, d**). Vertical distribution of anomalous T depicts warmer (cooler) environment at lower (upper) levels over the study region, which represents the destabilization of the tropospheric lapse rate. In contrast, wet spell composite profile of T' exhibits cooling in the lower levels and slight warming (< 0.5 K) in the upper troposphere. This upper level warming and lower level cooling in T' are not confined to single station but rather observed all over SE peninsular India. Although the vertical profile patterns of T' are consistent with the observations, magnitudes differ at certain pressure levels especially around boundary layer heights. For instance, boundary layer warming and cooling are stronger in ERA (black curve) than NCEP (red curve). The differences in the magnitudes of T and q will have a profound effect on the buoyancy of the parcels and CAPE, which will be discussed next.

Figure 10 depicts the percentage occurrence of CAPE, during dry and wet spells computed from observations and reanalysis products. An obvious over estimation of CAPE by reanalysis data compared to observations in both spells is noticed, and the values are more in MERRA and NCEP in dry spell and only in MERRA during wet spell, respectively. Although reanalysis products overestimate CAPE magnitude, they reproduce the differences in CAPE between spells reasonably well. Thus, it is clear from **Figure 10** that mean (or mode) of CAPE distribution is larger in wet spell than in dry spell. All reanalysis exhibits this feature, albeit with different magnitudes. For instance, difference between CAPE in wet and dry spells is nearly equal (~ 800 J kg^{-1}) when estimated with GPS and ERA-interim data. In contrast, NCEP- and MERRA-estimated mean CAPE values show smaller difference between spells (~250 J kg^{-1}

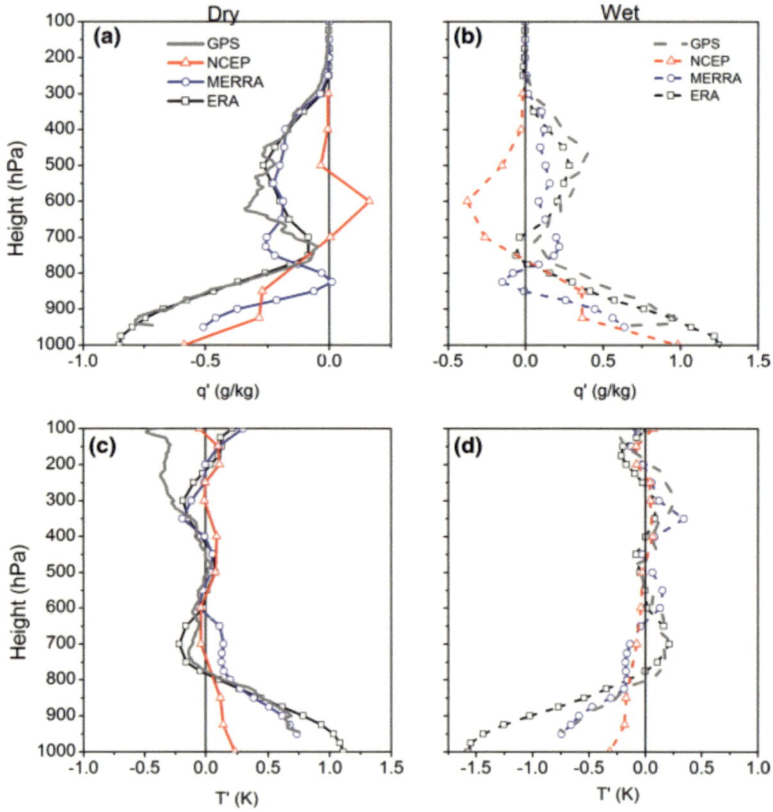

Figure 9.
Vertical profiles of anomalous moisture (a, b) and temperature (c, d) during dry and wet spells over Gadanki, from the sonde observations and reanalysis products (source [48]).

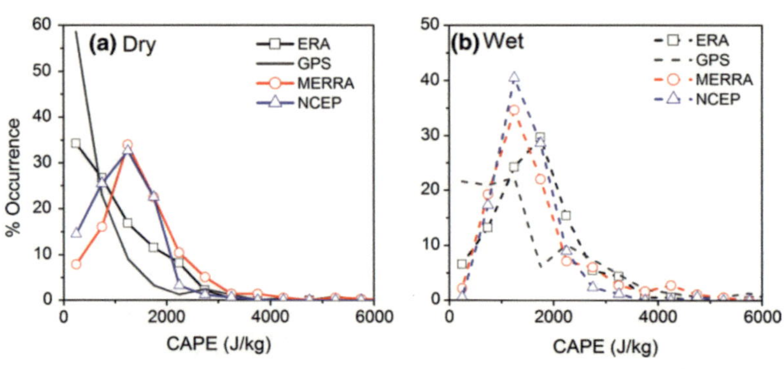

Figure 10.
Histograms of CAPE (at Gadanki) in (a) dry and (b) wet spells, respectively, as estimated by GPS sonde measurements and reanalysis outputs, illustrating the differences in CAPE between the spells (source [48]).

for NCEP and ~150 J kg^{-1} for MERRA). It is obvious from the above discussion that reanalysis products underestimate the CAPE variation within the spells when compared with sonde observations. Earlier, [34] have shown that while CAPE is strongly controlled by the properties of the boundary layer air, large positive buoyancy and realization of CAPE, however, occur above 600 hPa. Later, [47] also observed this feature over Gadanki using GPS radiosonde observations. Now, to examine how far reanalysis mimics this feature, parcel positive buoyancy area is divided into three layers (L1, 700–500 hPa; L2, 500–300 hPa; and L3, 300–200 hPa) to understand which positively buoyant layer is contributing more toward total CAPE in the spells. First, buoyancy for each layer is estimated from each sounding, and the contribution of each layer to total CAPE is estimated. These data are then grouped based on wet and dry spells, and the % occurrence for the layer contribution to total CAPE is estimated for each layer separately.

Figure 11 shows some intriguing similarities and differences in CAPE between spells, between layers, and between reanalysis and observations. In general, both reanalysis and observations show similar distribution of CAPE contributions by different positively buoyant layers. Among these layers, contribution of L2 to CAPE is more (~ 50%) in majority of the cases in both spells. From **Figure 11**, contribution of L1 to total CAPE is about ~38% (20–30%), and the remaining is by L3 in the wet (dry) spell. Also, the distribution in L1 during dry spell is relatively broader than in wet spell. Note that, for a few cases, the contribution of L1 (and to some extent L2) to CAPE is found to be as high as 100% in dry spell (**Figure 11a**), which indicates the vertical extent of clouds is limited in dry spell. This is consistent with report by [33] over Gadanki. All the reanalysis products reproduced above features; nevertheless, contributions of different positively buoyant layers to total CAPE are different. For example, the contribution of L3 to CAPE is relatively small in MERRA in wet spell. Although there are slight variations that exist in representation of

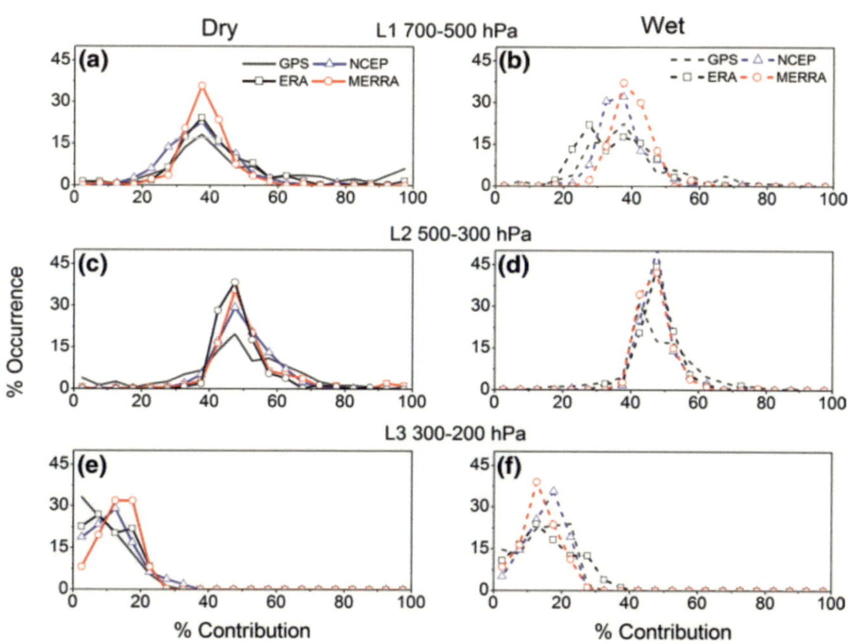

Figure 11.
Contribution of CAPE in different layers (L1: 700–500 hPa, L2: 500–300 hPa, L3: 300–200 hPa) to total CAPE in dry ((a), (c) and (e))wet ((b), (d) and (f)) spells at Gadanki. GPS sonde measurements and NCEP, ERA-interim, and MERRA reanalysis outputs are used in this analysis (source from [48]).

CAPE during the spells, it is clear from the analysis that ERA data reproduce similar CAPE variation between spells as obtained by radiosonde measurements.

Spatial variability of anomalous CAPE during wet and dry spells over SE peninsular India from the reanalysis products is shown in **Figure 12**. Note that, here, main focus is to find out the difference in dry (**Figure 12 (a)-(c)**) and wet (**Figure 12 (d)-(f)**) spells over SE peninsular India, and therefore, the analysis is restricted only to that region. It is clear from **Figure 12** that during wet spell, CAPE is larger in all the products [47]. These large CAPE values are not confined to single station but rather observed all over southeastern peninsular India. In contrast, negative CAPE values are seen during dry spell in all the reanalysis products. These –ve CAPE magnitudes, during dry spell, reconfirms that the thermal stability of the atmosphere during dry spell is significantly less for convection to trigger. Though there are similarities that exist among the reanalysis products, slight differences in magnitudes are noted especially during dry spell. For example, magnitudes of anomalous CAPE in MERRA are much less (~40 J kg^{-1}) than ERA (~200 J kg^{-1}). These differences are perhaps due to various convective parameterizations employed in the reanalysis products and also the vertical resolutions of the data products used.

Further, it is also noted from the buoyancy profiles during the spells that majority of positive buoyancy profiles show two peaks during dry spell and single upper tropospheric peak in wet spells at most of the grid points over southeast peninsular India (see Table 5 in [48]). Overall, majority of the buoyancy profiles, i.e., ~64%, exhibit bimodal distribution in dry spell, while single-peak buoyancy profiles are more prominent in wet spell (~56%), indicating the difference in the vertical distribution of parcel thermal buoyancy is not confined to single station but is a characteristic feature over the entire SE India. The analysis showed that, in this region, CAPE is higher in wet spell than in dry spell by ~1000 J kg^{-1}. Now it is imperative to understand the observed differences in CAPE between spells. In order to answer this, several plausible mechanisms are examined to explain the observed

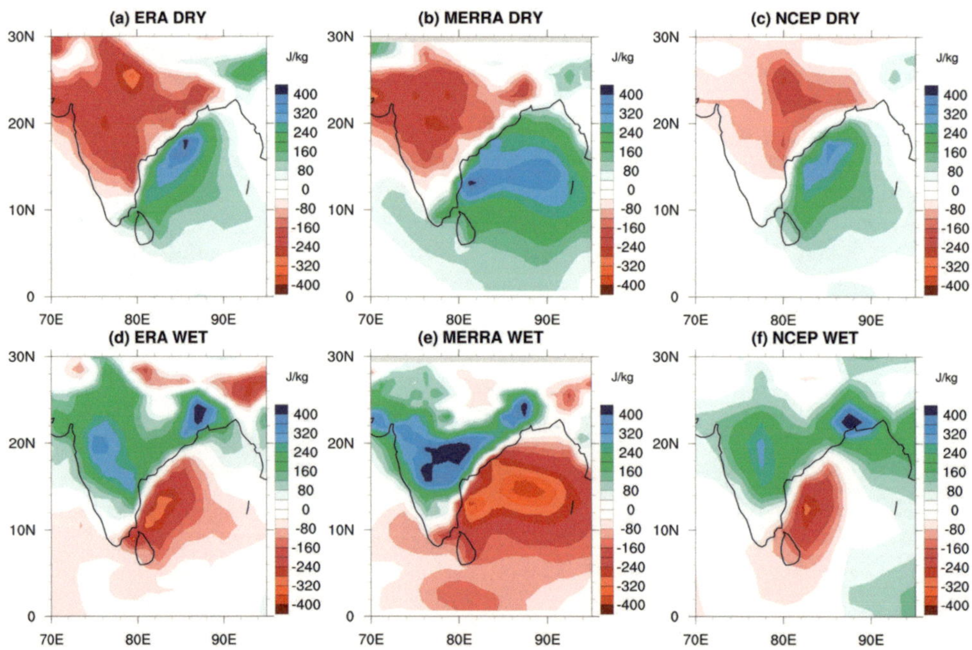

Figure 12.
Spatial distribution of anomalous CAPE in dry ((a)-(c)) dry and wet (d)-(f) spells, from three reanalysis products, over the southeast peninsular India (reproduced from the [48]).

CAPE differences between spells, i.e., time of sounding with respect to the rain occurrence, rapid rebuild-up of the instability, moistening of the atmosphere due to the evaporation of surface moisture in wet spell, enhanced low-level moisture convergence, enhanced downdrafts engendered by evaporative cooling, and drop dragging in dry spell. The weak CAPE in dry spell may not be sufficient to overshoot the frequent stable layers occurring in this spell. All these above mechanisms seem to be occurring in dry spell limiting the convection growth and CAPE [48].

5.3 Dynamical characteristics

In the previous sections characteristics of wet and dry spells are studied with respect to the thermodynamical point of view. The analysis reveals significant variability in surface moisture and temperature, vertical structure, and also CAPE between spells, which clearly indicates the thermal structure, available energy, and their forcing are different in spells over SE peninsular India. Therefore, one would expect differences in circulation features during wet and dry spells. Thus, in the present section differences in mean wind, vertical structure and diurnal variation with a special emphasis on monsoon quasi permanent systems (likes of LLJ and TEJ) over south eastern peninsular India are studied.

Majority of the data were obtained by collocated instruments available at Gadanki. Surface winds during the spells are measured from automatic weather station (AWS) along with Doppler sound detection and ranging (SODAR) wind components in the height region from 60 m to 1 km. Zonal and meridional winds above 1 km to upper troposphere (~18 km) are derived from low atmospheric wind profiler (LAWP) and Indian mesosphere-stratosphere and troposphere radar (IMSTR). Note that, although the instruments were not operated simultaneously as they were deployed in different years, it is believed that the mean winds represent the overall vertical structure and variability.

Figure 13 shows mean vertical profiles of mean wind and deviation during dry and wet spells of the monsoon over Gadanki. Since measurements utilized to generate this vertical wind structure by IMSTR, LAWP, SODAR, and AWS were not simultaneous, composite vertical profiles were not constructed. Rather, **Figure 13** shows average wind variation between spells in different altitude ranges (surface, obtained from AWS; 60 m to 1 km, obtained from SODAR; 600 m to 4 km, obtained from LAWP; and 3.6 to 19 km, obtained from IMSTR) [52].

Mean surface winds from AWS measurements (**Figure 13c**) during wet spells are relatively weaker than in dry spell, but the differences in zonal and meridional winds between the spells are not significant. However, the wind direction remained southwesterly in both spells. SODAR winds in the height region of 60 m–1 km remain northwesterly to westerly but exhibit large vertical variation in magnitude, particularly the zonal component. Zonal wind component attains their peak strength in the height region 200–500 m during both the spells, and in particular the intensity of zonal winds is stronger during dry spells (~4 ms^{-1}) than wet spell (<2 ms^{-1}), and further, the maximum difference between the spells is found in the height region of nocturnal low-level jet (NLLJ). In contrast, meridional winds are weak in amplitude (~1 ms^{-1}) during both the spells and do not show any significant differences between the spells. In the height regions from 600 m to 4 km, LAWP-derived winds continued to be westerly to northwesterly in both spells. The presence of LLJ is clearly seen in the zonal wind component, and interestingly, the height of LLJ peak varies during both spells (e.g., at 2.25 and 1.35 km in dry and wet spells, respectively). In addition, magnitude of the LLJ is also different with enhanced LLJ in the dry spell (16.8 ms^{-1}) than in wet spell (9.8 ms^{-1}). Interestingly, the difference in zonal wind between the spells is more pronounced above 1.5 km.

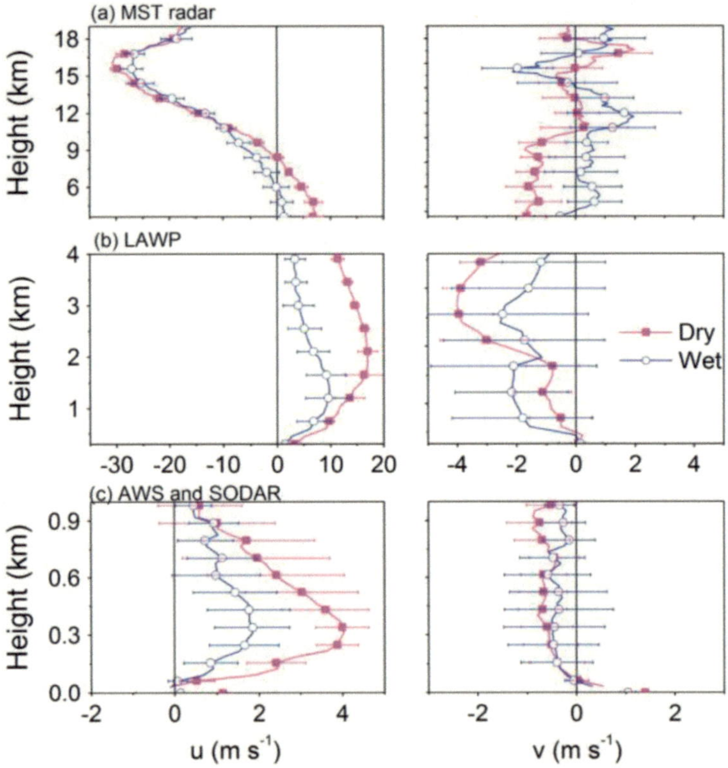

Figure 13.
Vertical profiles of (left column) zonal and (right column) meridional winds for wet and dry spells obtained from different instruments ((a) MST radar (1995–2013), (b) LAWP (1999–2000 and 2010–2011), and (c) SODAR (2007–2009) and AWS (2006–2013)). The average wind shown at 0 km (or surface) in ***Figure 13c*** *is obtained from AWS. Vertical profiles shown in* ***Figure 13a–c*** *are daily averages. The standard deviation is represented with error bars. Years in the brackets indicate data averaging periods (source from [52]).*

In contrary, the magnitude of meridional winds is relatively small and does not show significant variation between spells. IMSTR winds reveal that the vertical structure of wind in the height region of 3.6–19 km is different in both spells. There is a significant difference in vertical structure of winds from IMSTR, in the height region from 3.6 to 19 km, in both wet and dry spells (**Figure 13a**). These differences are pronounced in the lower and middle troposphere (below 8 km). The zonal wind profiles show typical summer monsoon circulation with low-level westerlies and strong upper tropospheric easterlies. Wind reversal height (i.e., from westerly to easterly), however, is different during both monsoon spells. The depth of westerlies is relatively shallow (deep) during the wet (dry) spell with zonal wind reversal occurring below 6 (8) km and then turns to easterly. On the other hand, the TEJ strength is found to be nearly the same in both spells with an average value of ~30 ms^{-1}. The height of the TEJ maximum is also found to be nearly the same during both spells (~16 km).

This section discusses the differences in the spatial variability of the jet streams between the spells. Composites of LLJ and TEJ for wet and dry spells and the wind anomaly (mean wind for dry spell-mean wind for wet spell) are estimated. Note that the main idea is to study the spatial variability of LLJ and TEJ when the monsoon convection is weak or active over southeast India. It allows us to examine the spatial extent of observed wind differences at Gadanki.

Figure 14 exhibits the spatial variation of mean zonal wind pattern for dry (**Figure 14a**) and wet (**Figure 14b**) spells and their difference (**Figure 14c**) on

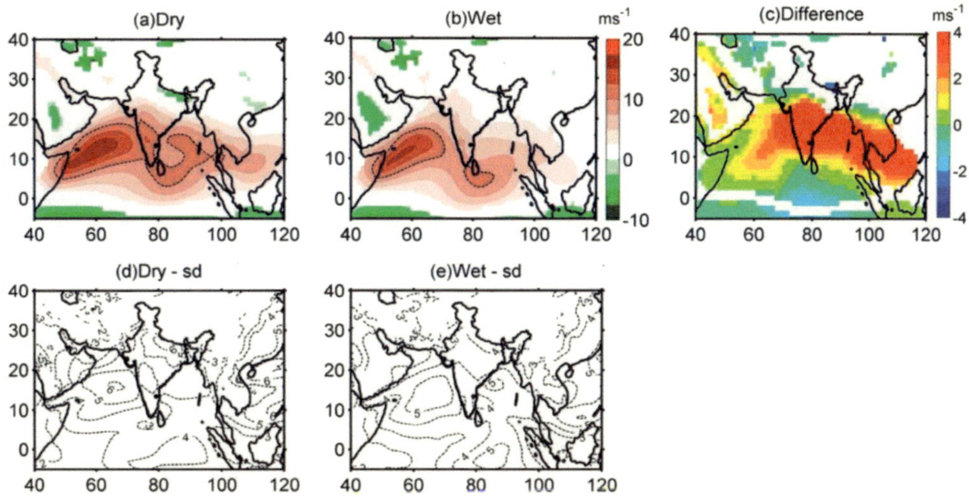

Figure 14.
Mean zonal wind on 850 hPa level from MERRA data for (a) dry and (b) wet spells, during 1995–2013, showing the spatial variation of LLJ. The black dashed line in ***Figure 4a*** *and* ***b*** *represents 10 ms^{-1} contour. (c) the difference in LLJ between spells (dry-wet). The dot denotes the location of Gadanki. (d and e) the spatial distribution of standard deviation of mean values for dry and wet spells, respectively (source from [52]).*

850 hPa level. Spatial distribution of the standard deviation of means for dry and wet spells is shown in **Figure 14d** and **e**, respectively. One commonality is the presence of LLJ in both the spells albeit with different magnitude and spatial distribution. During the dry spell (**Figure 14a**), the core of the LLJ splits over the Arabian Sea with one branch (say first branch) passing over the southern peninsular India centered around 16°N and the other toward southeast and veers cyclonically (second branch) before merging with the first branch in the Bay of Bengal (near the coast of Malay peninsula). On the other hand, only one branch (second branch) is present during the wet spell (**Figure 14b**). This splitting of LLJ in the Arabian Sea can be attributed to barotropic instability [49]. In this study, the splitting of LLJ is clearly evident during the dry spell (analogous to all-India active spell) [50, 51]. The presence of southward branch of LLJ during the break spell as reported by [48] is also seen here. In fact, this second branch is present in both spells with similar magnitude, as seen by the small wind anomaly present in that region (**Figure 14c**). **Figure 14c** clearly shows that large differences exist between the spells in LLJ magnitude and its spatial variation. A band of large positive wind anomaly passes over the Arabian Sea, Peninsular India, Bay of Bengal, and Malaysia with a maximum (~6 ms^{-1}) over the Southern Peninsular region. This large wind anomaly is significant and is occurring mainly due to the absence of first branch of LLJ during the wet spell. A negative anomaly in zonal wind is also observed in two regions, just south of the equator and near foot hills of the Himalayas. In general, the low-level westerly winds turn cyclonically in the North Bay of Bengal and become easterlies. These easterlies are clearly seen during the dry spell (or all-India active spell) (**Figure 14a**). As the monsoon trough moves northward to foot hills of the Himalayas during wet spell (or all-India break spell), the magnitude of easterlies became very weak (**Figure 14b**). In fact, the easterlies are completely absent over the Indian landmass during the wet spell.

Spatial variation of mean zonal wind and standard deviation during wet and dry spells along with zonal wind difference between the spells at 100 hPa level is described in **Figure 15**. The easterly winds are strong in both spells and seen

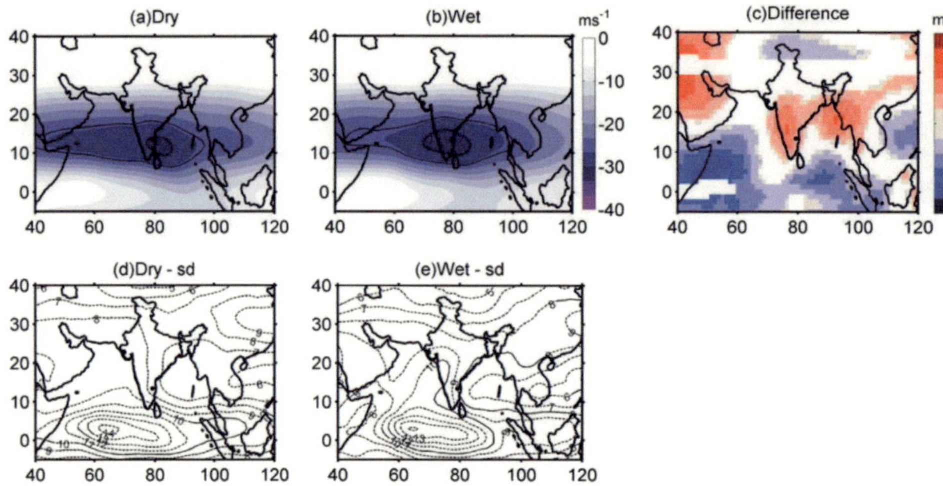

Figure 15.
*Same as **Figure 14** but for zonal winds at 100 hPa level, showing the TEJ variation. The black thin and thick solid lines in **Figure 15a** and **b** represent 24 and 28 ms^{-1} contours, respectively (source from [52]).*

prominently between 10°N and 20°N with winds as large as 30 ms^{-1} corroborating the radar observations made at Gadanki (**Figure 13a**). However, longitudinal extent of TEJ during dry spell is found to be more than in wet spell. It can be evidenced by 28 ms^{-1} (thick black solid line) and 26 ms^{-1} contour (thin black solid line) of TEJ. For instance, the 26 ms^{-1} contour is seen between the longitudes 40°E and 100°E during the dry spell, while it is only confined between 45 E and 95°E during the wet spell. Earlier, [49] observed a significant ISV in TEJ axis on 200 hPa level. They observed the axis of the TEJ along 15°N during the break spell, while a southward shift in the TEJ axis is observed during the active spell at 70°E. Such a north–south shift in TEJ axis is not observed in either of the spells here. The axis is found to be aligned along ~15°N latitude during both the spells.

6. Summary, conclusion, and discussion

Synthesis of spatial and vertical structures and regional characteristics of wet and dry spells over southeast peninsular India is presented in this chapter. In spite of its importance, not many reports exist on rainfall and its variation on different temporal scales over this region during southwest monsoon, partly because the rainfall in this region is relatively less, and it forms only a minor part of all-India rainfall (see **Figure 3a**). With the goal in mind, an attempt has been made to understand differences in thermal and dynamical characteristics and energetics of the atmosphere between wet and dry spells of the Indian summer monsoon over the southeast India. To better understand the aforementioned processes, data collected with a suite of unique instruments at NARL, model reanalysis data sets, and satellite rainfall products are effectively utilized.

It is observed that the difference in the thermal structure between wet and dry spells is significant only in the lower troposphere (<2–3 km). The distributions and mean CAPE values, which are measures of thermal instability, for wet and dry spells, are found to be different. Analysis indicates that mean CAPE in wet spell is found to be higher than that of in dry spell by ~1000 J kg^{-1}. The vertical extent of positive buoyancy profiles is deep during wet spells, while most of the buoyancy profiles during dry spells are limited in vertical extent. Further, the negative buoyancy areas are seen in most of the profiles during dry spell, and they are mostly centered on two height regions, ~700 and ~500 hPa. They, probably, are due to the strong inversions and stable layers present at those altitudes. The convection growth is limited in dry spells due to the presence of strong stable layers, weak CAPE, and a relatively dry environment.

Reanalysis data products indicate CAPE is higher in wet spell than in dry spell over the entire southeast peninsular India. Majority of buoyancy profiles show only one peak during wet spell, while bimodal vertical distribution is seen predominantly in dry spell. From this analysis it is found that the differences in CAPE and buoyancy vertical structure between wet and dry spells are not only confined to Gadanki but rather observed all over southeastern peninsular India, and they are characteristic features of wet and dry spells. Several possible mechanisms are invoked to explain observed CAPE differences between spells, i.e., rapid rebuild-up of the instability, moistening of the atmosphere due to the evaporation of surface moisture in wet spell, enhanced downdrafts engendered by evaporative cooling, and drop dragging in dry spell. The synthesis of all measurements and estimates indicate that the observed weak CAPE in dry spell may not be sufficient to overshoot the frequent stable layers occurring in this spell. Further, the strong (magnitude) and deep (in height) low-level wind shear observed in dry spell seems to be shearing apart the weakly buoyant updraft.

Diurnal variation of winds from surface to the lower stratosphere is studied during different spells of the monsoon. It is observed that, over the study region, the surface and low-level mean winds are stronger during dry spells. The surface wind (both zonal and meridional) exhibits a clear diurnal cycle with strong (weak) wind during day (night) in both spells. Both the amplitude and time of wind maxima change with the altitude. For instance, the zonal wind maxima observed at noon near the surface is shifted to early morning in the height region of 400 m–1.5 km (1 km) and then systematically to evening (noon) in the height region of 3–6 km (1–2.5 km) in dry (wet) spell. The depth of westerlies is deeper in dry spell than in wet spell. Also, the zonal wind reversal height shows clear diurnal variation in wet spell, while it is nearly the same in dry spell. The amplitude of the diurnal cycle increases with altitude up to 2 km and then decreases. Largest amplitudes of the diurnal cycle (>8 m s^{-1}) are found in the height region of 1–2 km. The splitting of LLJ into two branches over the Arabian Sea is quite pronounced in dry spell, with one branch passing over the peninsular India and the other branch veering cyclonically and joins the first branch in the Bay of Bengal. The strength and the axis of TEJ do not vary much between spells. These variations are compared and contrasted with earlier reports on jet streams.

The diagnostics made with the in situ observations and reanalysis products and the key results obtained can be exploited for the modeling purpose for better prediction of subseasonal variability of rainfall regionally.

Acknowledgements

The author would like to thank the director of the National Atmospheric Research Laboratory, Gadanki, for providing necessary facilities to carry out the work. The author would also like to thank Dr. T. Narayana Rao, NARL, for providing critical comment and helpful discussion in shaping out this work.

Acronyms

ITCZ	intertropical convergence zone
TCZ	tropical convergence zone

ISV	intraseasonal variability
ISMR	Indian summer monsoon rainfall
ISO	intraseasonal oscillation
IMD	India Meteorological Department
NARL	National Atmospheric Research Laboratory
MSL	mean sea level
GMT	Greenwich mean time
AWS	automatic weather station
CAPE	convective available potential energy
CINE	convective inhibition energy
LCL	lifting condensation level
LFC	level of free convection
EL	equilibrium level
ERA	European Center for Medium Range Weather Forecasting reanalysis
NCEP	National Center for Environmental Prediction
MERRA	Modern Era Retrospective analysis for Research and Applications
LLJ	Low-level jet
TEJ	tropical easterly jet
IMSTR	Indian mesosphere-troposphere stratosphere radar
LAWP	low atmospheric wind profiler
SODAR	sound detection and ranging

Author details

Mohana S. Thota
Department of Civil Infrastructure and Environmental Engineering,
Khalifa University, Abu Dhabi, United Arab Emirates

*Address all correspondence to: drmohanthota@gmail.com

References

[1] Ramage CS. Monsoon Meteorology (International Geophysics Series). Vol. 15. San Diego, California: Academic Press; 1971. 296 pp

[2] Slingo. Monsoon: Overview. UK: University of Reading, Elsevier Science Ltd.; 2003

[3] Pai DS, Rajeevan M. Summer monsoon onset over Kerala: New definition and prediction. Journal of Earth System Science. 2009;**118**:123-135

[4] Gadgil S, Kumar KR. The Asian monsoon—Agriculture and economy. In: Wang B, editor. The Asian Monsoon. Berlin: Springer; 2006. pp. 651-683

[5] Sikka DR, Gadgil S. On the maximum cloud zone and the ITCZ over Indian longitude during southwest monsoon. Monthly Weather Review. 1980;**108**:1840-1853

[6] Goswami BN, Ajay Mohan RS. Intra-seasonal oscillations and interannual variability of the Indian summer monsoon. Journal of Climate. 2001;**14**:1180-1198

[7] Goswami BN, Ajaymohan RS, Xavier PK, Sen gupta D. Clustering of synoptic activity by Indian summer monsoon intraseasonal oscillations. Geophysical Research Letters. 2003;**30**:1431. DOI: doi 10.1029/2002GL016734

[8] Jones C, Waliser DE, Lau WKM, Stern W. The Madden-Julian oscillation and its impact on northern hemisphere weather predictability. Monthly Weather Review. 2004a;**132**:1462-1471

[9] Singh SV, Kripalani RH. The south to north progression of rainfall anomalies across India during the summer monsoon season. Pure and Applied Geophysics. 1985;**123**:624-637

[10] Singh SV, Kripalani RH. Application of EEOF analysis to interrelationships and sequential evolutions of monsoon fields. Monthly Weather Review. 1986;**114**:1603-1610

[11] Singh SV, Kripalani RH. Low frequency intra-seasonal oscillation in Indian rainfall and outgoing long-wave radiation. Mausam. 1990;**41**:217-222

[12] Kripalani RH, Singh SV, Arkin PA. Large-scale features of rainfall and outgoing long-wave radiation over Indian and adjoining regions. Contributions to Atmospheric Physics. 1991;**64**:159-168

[13] Kriplani RH, Kulkarni A, Singh SV. Association of the Indian summer monsoon with the northern hemisphere midlatitude circulation. International Journal of Climatology. 1997;**17**:1055-1067

[14] Kulkarni A, Sabade SS, Kriplani RH. Spatial variability of intra-seasonal oscillations during extreme Indian monsoons. International Journal of Climatology. 2009;**29**:1945-1955

[15] Webster PJ, Magana VO, Palmer TN, Shukla J, Tomas RA. Monsoons: Processes, predictability, and the prospects for prediction. Journal of Geophysical Research. 1998;**103**:4451-4510

[16] Gadgil S. The Indian monsoon and its variability. Annual Review of Earth and Planetary Sciences. 2003;**31**:429-467

[17] Rajeevan M, Bhate J, Kale JD, Lal B. High resolution daily gridded rainfall data for the Indian region: Analysis of break and active monsoon spells. Current Science. 2006;**91**:296-306

[18] Swaminathan MS. Abnormal monsoon and economic consequences:

The Indian experience. In: Fein JS, Stephens PL, editors. Monsoons. Washington, D.C: John Wiley & Sons; 1987. pp. 121-133

[19] Parthasarathy B, Munot AA, Kothawale DR. Regression model for estimation of Indian food grain production from summer monsoon rainfall. Agricultural and Forest Meteorology. 1988;**42**:167-182

[20] Gadgil S, Abrol YP, Seshagiri Rao PR. On growth and fluctuation of Indian food grain production. Current Science. 1999a;**76**:548-556

[21] Abrol IP. India's agriculture scenario. In: Abrol YP, Gadgil S, Pant GB, editors. Climate Variability and Agriculture. New Delhi: Narosa Publishing House; 1996. pp. 19-25

[22] Blanford HF. Rainfall of India. Monsoon Monograph—India Meteorological Department. 1886;**2**:217-448

[23] Ramamurthy K. Monsoon of India: Some aspects of the 'break' in the Indian southwest monsoon during July and August. Forecasting Manual, India Meteorological Department, Pune, India. 1969;**18**(3-IV):1-57

[24] Magna V, Webster PJ. Atmospheric circulations during active and break periods of the Asian monsoon. In: Preprints of the Eighth Conference on the Global Ocean-Atmosphere-Land System (GOALS). Atlanta, GA: Amer. Meteorol. Soc.; 1996

[25] De US, Lele RR, Natu JC. Breaks in southwest monsoon. Report No. 1998/3. New Delhi, India: India Meteorological Department; 1998

[26] Krishnan P, Kunhikrishnan PK, Nair SM, Ravindran S, Jain AR, Kozu T. Atmospheric boundary layer observations over Gadanki using lower atmospheric wind profiler-preliminary results. Current Science. 2003;**85**:75-79

[27] Gadgil S, Joseph PV. On the breaks of the Indian monsoon. Proceedings of the Indian Academy of Sciences - Earth & Planetary Sciences. 2003;**112**:529-558

[28] Ramesh Kumar MR, Desai P, Uma R. A new criterion for identifying breaks in monsoon conditions over Indian subcontinent. Geophysical Research Letters. 2004;**31**:L18201. DOI: 10.1029/2004GL020787

[29] Prasad VS, Hayashi T. Large-scale summer monsoon rainfall over India and its relation to 850 hPa wind shear. Hydrological Processes. 2007;**21**:1992-1996

[30] Klingaman NP, Weller H, Slingo JM, Innes PM. The intraseasonal variability of the Indian summer monsoon using TMI sea surface temperatures and ECMWF reanalysis. International Journal of Climatology. 2008;**21**:2519-2539

[31] Krishnamurthy V, Shukla J. Intraseasonal and interannual variability of rainfall over India. Journal of Climate. 2000;**13**:4366-4377

[32] Rajeevan M, Gadgil S, Bhate J. Active and break spells of the Indian summer monsoon. Journal of Earth System Science. 2010;**119**:229-247

[33] Rao TN, Uma KN, Satyanarayana TM, Rao DN. Differences in draft core statistics from wet spell to dry spell over Gadanki (13.5° N, 79.2° E), India. Monthly Weather Review. 2009;**137**:4293-4306

[34] McBride JL, Frank WM. Relationships between stability and monsoon convection. Journal of the Atmospheric Sciences. 1999;**56**:24-36

[35] Bhat GS, Chakraborty A, Nanjundiah RS, Srinivasan J. Vertical thermal structure of the active and weak phases of convection over the north Bay of Bengal: Observation and model results. Current Science. 2002;**83**:296-302

[36] Webster PJ et al. The JASMINE pilot study. Bulletin of the American Meteorological Society. 2002;**83**:1603-1630

[37] Yi L, Lim H. Comparison of convective environments in tropical and extratropical atmospheres with 1989-2002 radiosonde data. In: 26th Conference on Hurricanes and Tropical Meteorology. AMS; 2004. pp. 1-71

[38] Bhat GS et al. BOBMEX: The Bay of Bengal monsoon experiment. Bulletin of the American Meteorological Society. 2001;**82**:2217-2243

[39] Mapes BE, Houze RA Jr. An integrated view of the1987 Australian monsoon and its mesoscale convective systems. Part I: Horizontal structure. Quarterly Journal of the Royal Meteorological Society. 1992;**118**:927-963

[40] Williams E, Renno N. An analysis of conditional instability of the tropical atmosphere. Monthly Weather Review. 1993;**121**:21-36

[41] Cifelli R, Rutledge SA. Vertical motion, diabatic heating, and rainfall characteristics in N. Australia convective systems. Quarterly Journal of the Royal Meteorological Society. 1998;**124**:1133-1162

[42] Halverson JB, Rickenbach T, Roy B, Pierce H, Williams E. Environmental characteristics of convective systems during TRMM-LBA. Monthly Weather Review. 2002;**139**:1493 1493-1493 1509

[43] Johnson RH, Ciesielski PE, Hart KA. Tropical inversions near the 0 °C level. Journal of the Atmospheric Sciences. 1996;**53**:1838-1855

[44] Dee DP, Uppala SM, Simmons AJ, Berrisford P, Poli P, Kobayashi S, et al. The ERA-interim reanalysis: Configuration and performance of the data assimilation system. Quarterly Journal of the Royal Meteorological Society. 2011;**137**(656):553-597

[45] Kalney E et al. The NCEP-NCAR 40-year reanalysis project. Bulletin of the American Meteorological Society. 1996;**77**:437-471

[46] Rienecker MM, Suarez MJ, Gelaro R, Todling R, Bacmeister J, Liu E, et al. MERRA: NASA's modern-era retrospective analysis for research and applications. Journal of Climate. 2011;**24**:3624-3648. DOI: 10.1175/JCLI-D-11-00015.1

[47] Mohan TS, Rao TN. Variability of the thermal structure of the atmosphere during wet and dry spells over Southeast India and its implication on draft cores. Quarterly Journal of the Royal Meteorological Society. 2012;**138**:1839-1851

[48] Mohan TS, Rao TN, Rajeevan M. Differences in CAPE between wet and dry spells of the monsoon over the southeastern peninsular India. Meteorology and Atmospheric Physics. 2018. DOI: 10.1007/s00703-018-0590-9

[49] Findlater J. Mean monthly airflow at low-levels over the western Indian Ocean. Geophysical Memoirs. 1971;**16**:1-53

[50] Joseph PV, Sijikumar S. Intraseasonal variability of the low-level jet stream of the Asian summer monsoon. Journal of Climate. 2004;**17**:1449-1458

[51] Sathiyamoorthy V, Pal PK, Joshi PC. Intraseasonal variability of the tropical easterly jet. Meteorology and Atmospheric Physics. 2007. DOI: 10.1007/s00703-006-0214-7

[52] Mohan TS, Rao TN. Differences in mean wind and its diurnal variations between wet and dry spells of the monsoon over southeast peninsular India. Journal of Geophysical Research. 2016. DOI: 10.1002/2015JD024704

Chapter 8

The Disturbed Habitat and Its Effects on the Animal Population

Maria Teresa Capucchio, Elena Colombino, Martina Tarantola, Davide Biagini, Loris Giovanni Alborali, Antonio Marco Maisano, Federico Scali, Federica Raspa, Emanuela Valle, Ilaria Biasato, Achille Schiavone, Cristian Salogni, Valentina Bar, Claudia Gili and Franco Guarda

Abstract

Changes in the "habitat" may interfere with the normal functioning of all biological systems. The existence of relationships between environmental changes and health in humans and animal species is well known and it has become generally accepted that poor health affects the animal's natural behaviors and animal welfare and, consequently, food safety and animal production quality. Microclimate alterations, husbandry-management conditions, quality of human-animal interactions, feeding systems, and rearing environment represent the main factors that could negatively affect animal welfare and may produce behavioral, biochemical, endocrine, and pathological modifications in domestic and wild animals. Particularly, high stress levels can reduce the immune system response and promote infectious diseases. Adverse socio-environmental factors can represent a major stimulus to the development of different pathologies. This chapter will discuss the main pathological modifications described in domestic and wild animals due to "disturbed habitat" paying more attention to critical points detected in standard breeding systems.

Keywords: disturbed habitat, pathology, farm/zoo and wild animals, microclimate, housing systems, human-animal interactions, social interaction

1. Introduction

Disturbance has been defined as "a change in conditions which interferes with the normal functioning of a biological system" [1]. A "disturbed habitat" is an ecological concept indicating a temporary change in environmental conditions, which causes a pronounced change in the ecosystem. Disturbances can be human-caused or natural. Human disturbances include plowing, digging, construction activities, mowing, spraying weed-killing chemicals, clearing land for a garden, burning, severe livestock overgrazing, and so on. Natural disturbances include lightning strikes and fire; temperature changes, strong winds, ice storms, and tornadoes that topple or damage trees, heavy rain, flooding, hail, and erosion; and drought and earthquakes.

The existence of relationships between the changes in the environment and health is well known, and it has been documented by numerous scientific studies conducted over the past half century within all animal species and humans. Moreover, it has become generally accepted that poor health conditions can produce behavioral alterations and consequently affect the quality and safety of animal products.

In the recent years, consumers paid a great attention to the health and welfare of reared animal species. In 2007, in the article 13 of the Lisbon Treaty [2], the European Union has recognized animals as "sentient beings," capable of feeling pleasure and pain. The OIE code recognizes the "Five Freedoms" as valuable guidance in farm animal welfare/health [3]. The Five Freedoms concept analyzes the main domains related to the raising and handling of animals like feeding/nutrition, housing, health, and behavior, and it is used by various animal welfare standards to assess the animals' conditions. They were formulated in the early 1990s and are now well recognized as highly influential in the animal welfare arena. However, a marked increase in scientific understanding over the last two decades shows that the Five Freedoms do not capture, either in the specifics or the generality of their expression, the breadth and depth of current knowledge of the biological processes that are germane to understanding animal welfare and to guiding its management [4].

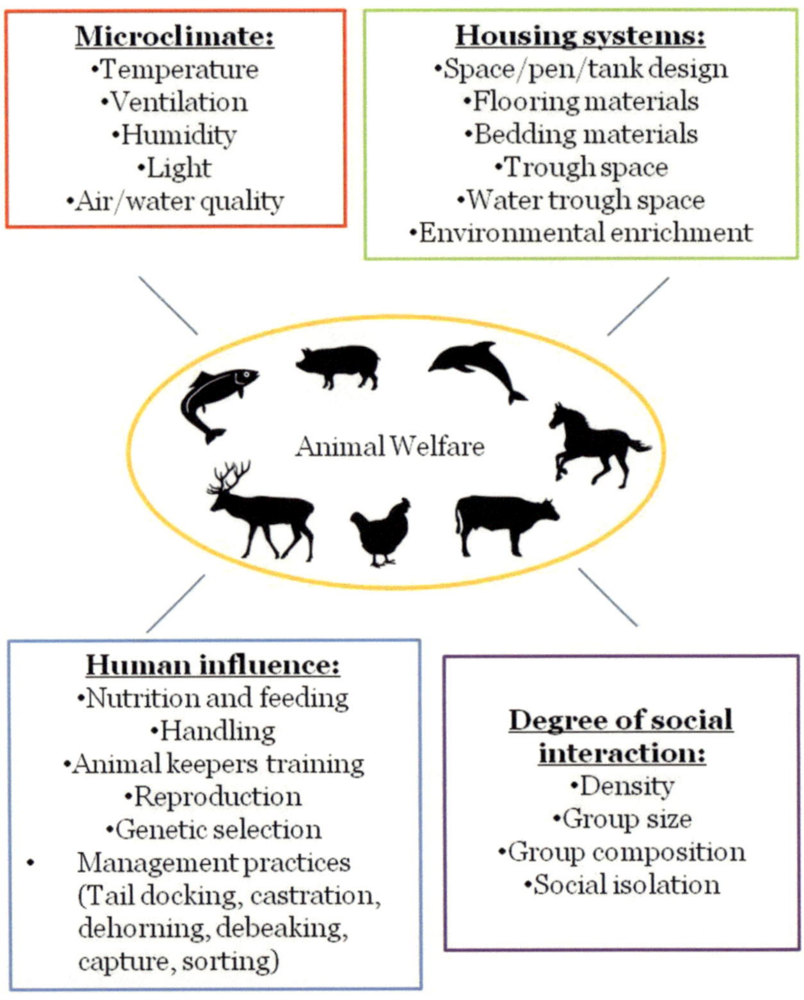

Figure 1.
Main critical points for animal welfare.

In fact, welfare criteria and parameters vary among and within species and depend directly on human utilitarian interests about them. The establishment and governance of animal care procedures impact directly on welfare of every managed species and should be carried out with "human care," thus anticipating and precisely estimating the resources to be provided to the animals by humans.

Advocacy groups often claim that animals in human care "only" deserve a good life "worth living." For animals to have "lives worth living," it is necessary, overall, to minimize their negative experiences and at the same time to provide them with opportunities to have positive experiences. To ensure this, during the last 10–20 years, national and international regulations or codes of welfare have increasingly included provisions that extend the welfare management focus to include elements well beyond the basic survival needs of farm animals [4].

Stressing factors due to an unsuitable habitat may produce behavioral, biochemical, and endocrine modifications in all the individuals that may be monitored by a series of well-known stress indicators such as the hematological profile, adrenal hormones (cortisol and its metabolites), acute-phase proteins, and d-ROMs. Biochemical modifications may lead to morphological alterations clinically manifested or not.

This chapter will discuss the pathological modifications affecting farm, zoo, and wild animals due to "disturbed habitat" addressing specific critical points (**Figure 1**) detected in standard breeding systems (**Figure 2**) for farm animals and species living in natural/semi-natural habitat for zoo/wild animals. Sections have been organized according to the following division: farm animals (cattle and small ruminants, pigs, equine species, poultry, and fish), zoo, and wild species.

Figure 2.
Intensive livestock farming: (A) cattle, (B) pig, (C) horses, and (D) poultry.

2. Farm animals

2.1 Cattle and small ruminants

Ruminants belong to the order Cetartiodactyla, which encompasses numerous species, and only a minority has been domesticated including cattle, sheep, and goats. Although these are suited to different habitats, in intensive farming systems, domestication has led to exposure to different stressors potentially responsible of pathologies.

For centuries, cattle have been grown in a traditional manner, within small farms, mainly grazing. Since the second half of the nineteenth century, the continuous demand of protein products and the availability of grains and protein sources to low costs led to an intensive, highly specialized production system, where animals are "adapted" to meet the constraints caused by their housing conditions and the management practices [5], thus restricting their natural behaviors. Furthermore, individual selection for enhanced production traits has placed an even greater metabolic demand on these animals.

The microenvironment experienced by cattle in houses, on open feedlots or at pasture is determined by the *microclimate*. Beef cattle can tolerate and adapt to a wide range of air temperatures, and metabolic heat production increases with increasing feed intake. Microclimate changes (e.g., inadequate ventilation, extreme temperatures, high relative humidity, ammonia concentration, etc.), affect the animal's immune response resulting in **respiratory and enteric diseases**, the major welfare problems in beef cattle [6].

The *housing system* could play an important role in cattle welfare [7]. Loose housing systems allow more freedom of movement than tether systems, also offering the animals the possibility of experiencing more natural social behaviors. The resting area is one of the most important areas, especially in a cow facility. Lying down is a basic requirement, and repeated deprivation is aversive to cows. Lying times are lower and standing times are higher when dairy cattle are forced to use hard surfaces. Particularly, in dairy cow, the poor hygiene and the materials of the bedding leads to **udder problems**, as manure may compromise cow comfort and increase the risk of intramammary infections. The type of flooring on which animals walk has been found to affect their welfare by impairing locomotion and increasing the occurrence of **hoof disorders and lameness**, which represent a major concern for the dairy industry because it negatively affects milk production. Beef cattle kept on slatted floors show a higher incidence of abnormal standing and lying movements and also a higher incidence of injuries than animals kept on concrete floor with fully or partially straw-bedded areas. A long duration of grazing periods, associated with frequent manure removal during the housing period, is probably a key factor for limiting the occurrence of podal lesions.

As far as *social interactions* are concerned, mixing and regrouping of cattle increase the incidence of agonistic behaviors and have also disadvantages from a health perspective. Older and more aggressive animals may cause trauma and continuous and severe stress to lower ranking calves (bullers). Small and young animals are more prone to diseases if kept with larger and older animals. For these reasons, groups should be made up with animals of similar age, weight, and sex [5]. Moreover, overcrowding and the reduced space at the manger are one of the most critical factors negatively affecting cattle welfare by increasing competition among pen-mates, causing the buller steer syndrome, decreasing the feed intake, reducing the time spent resting, eating, and ruminating, and increasing lesions, such as trauma on bones and joints, osteoarthropathies, prepuce injury, and tail-tip necrosis. In most intensive farming systems, the separation of the dairy calves from their mother in the period immediately after birth may have negative consequences for the health and welfare of cows and calves. Particularly, the socialization of calves may profit from staying with the dam, preferentially in a group [5].

Husbandry practices can have a tremendous effect on cattle provoking an increase in the prevalence of stress responses and physical injuries [8]. In fact, a positive attitude of the stockperson in handling and taking care of the animals seems to improve cattle welfare. The age of the farmers is also responsible for the less efficient management and consequently poor welfare of the animals. Not well-trained milkers may produce **teats injuries** that predispose to mastitis.

Furthermore, the welfare of any animal clearly depends on the provision of sufficient food to supply principally energy (Net Energy [NE]), proteins, amino acids, fatty acids, minerals, and vitamins, which are essential for the functions of life (maintenance, growth, activity, and reproduction). Failure to provide sufficient NE and optimal amounts of specific nutrients can lead to severe loss of body condition, infertility, and severe metabolic disorders. Growing beef cattle, housed, yarded or on feedlots, and presented with high energy and low fiber rations *ad libitum* are at risk of **digestive disorders** (**Figure 3A**). The most common of these ones is subacute ruminal acidosis, which occurs when the fermentation rate and hence the volatile fatty acid production exceed the buffering capacity of the rumen, but it is possible to observe also fatty liver, ketosis, displaced abomasum, liver abscesses, and laminitis. Unnatural foraging regimes, possibly exacerbated by restrictive environments, are thought to elicit stereotypic oral behavior in cattle, such as tongue-rolling, object-licking, chain-chewing, or bar-biting [6].

For all the reasons stated above, the authors hypothesize that the stress related to the intensive livestock farming could also represent a mechanotransduction-promoting factor of subclinical pathological changes such as **coronary arteriosclerosis** (**Figure 3B**), which has been frequently reported at slaughterhouse in both calves and beef cattle [9].

Basically, the major farming systems of small ruminants are those based on pasture (extensive-grazing), the indoor ones (intensive-industrial), and the semi-intensive. The negative impact of intensification of breeding systems can be observed at several levels and is very similar to what has been discussed above for the cattle. However, limited studies on the small ruminant welfare have been carried out, since they are considered very rustic animals able to cope with prohibitive environmental conditions and inadequate management practices, without harming their welfare and productive performances. This aspect has been overrated for many years considering that also in sheep and goats, stress can impair growth rate, wool growth, and feed conversion efficiency, also leading to the development of multi-factorial diseases such as **mastitis, laminitis, and metabolic disorders,** and increasing the frequency of abnormal behaviors (aggressive behavior), stereotypies, and vocalizations [10].

The *microclimate* is fundamental in preventing **respiratory diseases**. Indeed, animals allocated in hot and dusty environments are more prone to develop bacterial or viral pneumonia. Additional stressors could be found in the extensive systems, such as climatic extremes, that may evoke a **decrease in feed intake efficiency** and utilization, disturbances in water, protein, energy, and mineral balances, enzymatic reactions, hormonal secretions, and blood metabolites.

The *housing system* is fundamental for small ruminant welfare too: only few animals are reared in extensive production systems in which animals are free to move and perform their physiological and behavioral functions; most of them are housed only during the night and in the periods when grazing is not feasible. In any case, it is fundamental to understand that maintenance of good hygiene conditions, correct dimensioning of structural parameters, and adoption of proper management practices are important in either type of system.

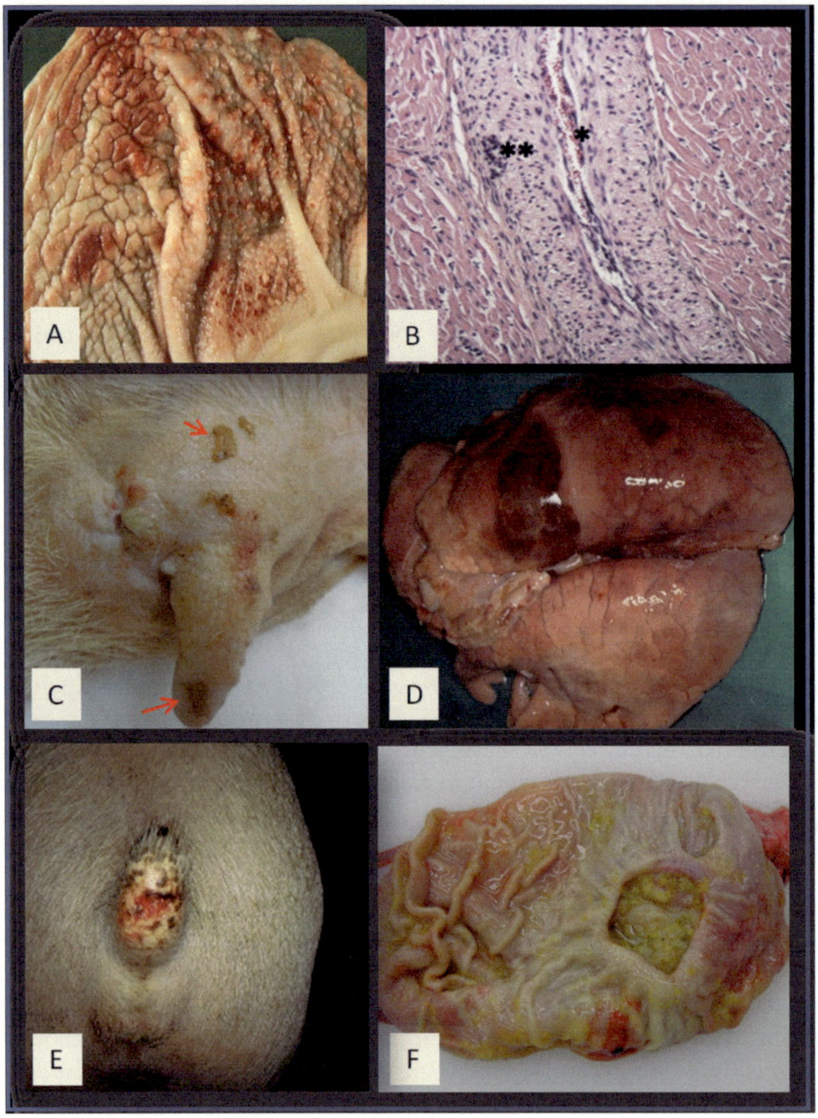

Figure 3.
(A) Beef cattle—abomasitis due to improper nutrition. (B) Beef cattle—heart. Arteriosclerotic alterations: diffuse intimal hyperplasia (*), medial smooth muscle cells reoriented and with disseminated vacuolar degeneration of the cytoplasm, moderate medial hypertrophy/hyperplasia (**). Histological section stained with H & E. Original magnification 20×. (C) Goat—udder. Traumatic teat injuries caused by milking. (D) Pig—lung. Pulmonary sequestration due to Actinobacillus pleuropneumoniae infection. (E) Pig—tail-biting lesions. (F) Pig—gastric ulcer due to improper nutrition.

With regard to *social interactions*, separation of goats/sheep from the group and re-introduction (e.g., for shearing or milking) and introduction of goats/sheep into established groups are stressful. One measure to reduce the effects of separation and re-introduction is to enable the (separated) goat/sheep to still hear and smell the other goats.

As seen above for cattle, *human-animal interaction* is a key factor also in the welfare of small ruminants too, and it is not unusual to find shepherds who have no specific skills or are not aware of the welfare standards of the animals [3]. An inadequate milking may produce teat injuries (**Figure 3C**) which is why specific training of farm crews should therefore be encouraged. Finally, an inadequate pasture in terms of quality and quantity can lead to **nutritional unbalance** with **liver disease, enzootic ataxia, pregnancy toxemia, hypocalcaemia, diarrhea,** and **enterotoxaemia**.

2.2 Pigs

Genetic selection in domestic pigs has been widely exploited in order to achieve specific phenotypic characteristics. Many pigs are raised in intensive conditions and thus strongly conditioned by the environment where they live. Moreover, even free-ranging domestic pigs and wild boars may be strongly influenced by human activities. Several signs of suffering in swine have been described and they can be quantified using animal-based measures (ABMs) [11, 12]. Furthermore, researches on pig welfare and ABMs led to the identification of "iceberg indicators" such as body injuries and ear and tail lesions. These indicators can be a proxy of "disturbed habitat" which is strongly influenced by microclimate and/or management. *Microclimate* heavily affects the stress conditions for pigs, particularly in intensive farming where different age groups require different microclimate standards (air, temperature, and humidity). If ventilation and air quality are not optimal, respiratory disorders, such as pneumonia (**Figure 3D**) and/or pleuritis from opportunistic pathogens, may occur, thus increasing the mortality. Variations in temperature and humidity (outside the thermal comfort) result in abnormal behaviors. For example, distressed pigs show increased huddling due to excessive cold weather and panting due to excessive hot weather [11]. Proper *management* is the key to maintain suitable habitat conditions for both intensive and extensive pig farming. Housing systems affect both animal behavior and physical conditions. In the intensive farming, floor types (e.g., slatted or solid), space allowance, and availability of bedding material influence incidence of bursitis, erosions, lameness, and shoulder ulcers. Moreover, the type of flooring directly affects the hygiene of the pig's body and the risk of developing enteric disorders. In the extensive farming, pigs must always have access to proper shelters; otherwise, outbreaks of severe enteric and respiratory disorders will occur increasing also the mortality rating. Appropriate structures and adequate space allowance for activities such as resting, feeding, and drinking are directly related to social behavior and interactions. Indeed, the environment in which pigs are confined influences the *degree of social interactions*. When a new group of pigs is formed, a stable social hierarchy is usually established in 1 or 2 days. During this initial phase, negative interactions arise and their outcomes may be observed mainly as wounds on the body. Once the hierarchy is established, negative interactions drastically subside while positive interactions (e.g., grooming, sniffing, nosing, and liking) and exploratory behaviors become prevalent [12, 13]. Nevertheless, rearing conditions typical of intensive housing systems can exacerbate inappropriate behaviors such as stereotypies (e.g., sham chewing) and negative interactions (e.g., ear and tail biting) (**Figure 3E**) [14]. *Human influence* on genetic selection and daily management is one of the most important variables which can exacerbate consequences of "disturbed habitat." Indeed, daily management errors, such as improper nutrition or feeding, may lead to severe conditions like gastric ulcers (**Figure 3F**) or toxic states (e.g., salt poisoning) which can cause high mortality [15]. Clear differences in the body condition scores of pigs of the same age are also a direct consequence of inadequate feeding. Genetic selection has led to great production results improving parameters such as reproductive performances, meat production, daily weight gain, and feed conversion ratio. However, this intense selection has made pigs less able to adapt to certain environmental situations (e.g., thickness of the subcutaneous fat layer), with organs at the limit of physiological potentiality (e.g., heart), leading to an increased risk of pathologies such as hernias and mulberry heart disease [16]. Pigs are also selected to be more prolific but, without adequate assistance, there is a drastic increase of newborn piglet mortality. Finally, human influence on pig management has repercussions on infectious diseases, which negatively affect pig health, such as colibacillosis, polyserositis, enzootic pneumonia, post weaning multisystemic wasting syndrome, and porcine reproductive and respiratory syndrome.

2.3 Equines for meat production

Equine meat consumption depends on cultural and traditional customs. Since it is a niche product, scientific community has made little efforts to define the main factors responsible for a "disturbed habitat" in this category. Equine breeds specifically selected for meat production do not exist and genetic selection focused more upon preserving and improving traits related to horses' morphology and performance. Therefore, although equines' domestication dates back to 5000 years ago, these species still retain the ancestral characteristics of their progenitors and feral equine populations can provide information about many aspects of equine behavior (e.g., social and foraging behavior). Considering the most important *microclimate factors* that negatively affect the equine habitat, insufficient ventilation and inadequate air quality may cause an increased exposure to gaseous ammonia and airborne dust that contain high levels of organic particulates including mite debris, microbes and vegetative material with varying content of endotoxins. The inhalant exposure to those irritant factors is implicated in the pathogenesis of **chronic inflammatory pulmonary disorders** such as inflammatory airway diseases (IAD) and recurrent airway obstruction (RAO) [17].

Equines reared for meat production are housed in conditions that markedly differ from those in which they evolved. As a consequence, those animals attempt to adapt to the conditions in which they are kept performing functionless and repetitive activity known as **stereotypic behaviors** that include crib-biting, wind sucking, box walking, and weaving [18]. Therefore, using equine stereotypies as welfare indicators should lead to perform management changes to enhance equine's welfare.

Bedding is an essential component in the *housing of the equine stabled. Bedding* should be dry, clean and not dusty, providing comfort, allowing animals to express their natural behavior of lying and resting and also avoiding the risk of **hoofs and skin lesions** [19].

Regarding the equines' opportunity to perform normal behaviors, it is important to guarantee an adequate space allowance to prevent aggressive reactions that might lead to stress competition for resources and for hierarchy establishment with consequent **physical injuries**. Indeed, wild horses live in relatively stable harem bands, so the overcrowding and the high rates of regrouping of intensively farmed horse may cause an increase in aggressiveness and injuries [20]. On the contrary, in nature, donkeys adapt easily, and their social organization depends on the availability of food and water resources. Therefore, the competition in the stabled donkeys, probably, could be increased if the available resources are not accessible to all, but their behaviors in farmed conditions need to be further studied.

Equines are grazing herbivores adapted to eating a forage-based diet. In nature, horses and donkeys spend about 16 hours per day foraging over wide distances, and this is essential for the health of their gastrointestinal tract and for their behavioral needs. On the contrary, equines in the breeding farms are fed high-energy, low-fiber concentrates, and this lack of foraging opportunity along with the high amount of concentrate feedstuff has been directly linked to the onset of **gastrointestinal disorders such as** gastric ulcerations (**Figure 4A**) and colic [19]. Equine gastric ulcer syndrome (EGUS) is reported in domesticated horses mainly involved in athletic endeavors. EGUS prevalence and severity have been correlated with the type of training and management practice. Common known risk factors have been identified in intense exercise, high grain-low roughage diet, water deprivation, fasting, hospitalization, and overdose of NSAIDs. In particular, excessive ingestion

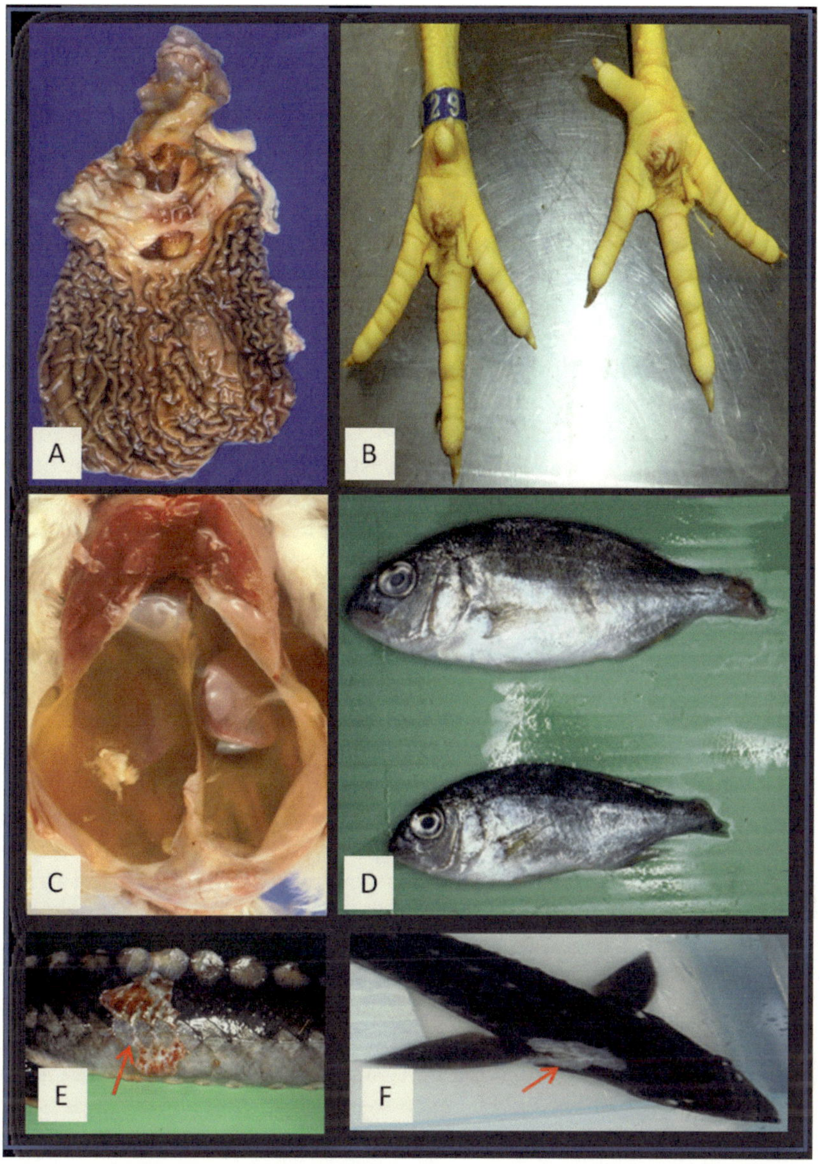

Figure 4.
(A) Horse—gastric ulcer due to improper nutrition. (B) Chicken—footpad dermatitis. (C) Chicken—ascites. (D) Sea bream (Sparus aurata)—peduncle mutilation caused by bite in overcrowding breeding conditions. (E) Sturgeon (Acipenser spp.)—skin erosion caused by an inappropriate and traumatic manipulation. (F) Sturgeon (Acipenser spp.)—dark color and a skin whitish patch due to Flavobacterium spp. infection. Evidence of stress and opportunistic infections.

of carbohydrates causes a rapid proliferation of the hindgut gram-positive bacteria *Streptococcus bovis* and *Streptococcus lutetiensis* that lead to very acidic conditions with pH lower than 4. Low pH in the large intestine causes the death and lysis of a large number of bacteria and the release of the toxic components which are absorbed from the gut into the bloodstream and may cause the development of **laminitis** [21].

2.4 Poultry

The concept of "disturbed habitat" in poultry farming can be almost entirely related to the continuously increasing production levels of the breeding programs, which are focused on increasing the growth rates and decreasing the feed conversion ratios of the animals. These *management procedures* lead to remarkable imbalances between the high potential productivity of birds (as a result of the targeted genetic selection) and their ability to physiologically adapt. These imbalances are frequently associated with homeostatic dysregulation and pathological changes of organs that supply the energy for production and maintenance (liver and cardiovascular system) or muscle tissue severely forced to obtain a fast weight increase. The other "disturbed habitat" conditions may strictly depend on microclimate alterations or management defects related to housing system and degree of social interactions.

With regard to the *microclimate* alterations, heat stress is the most common physical environmental stressor that can lead to alterations in the intestinal epithelium integrity and microbiota composition (with development of necrotic enteritis), hyperthermia, heat exhaustion, and death [22, 23]. Multiple behavioral, physiological, and health issues, such as reduced feed consumption, neuroendocrine disorders, electrolyte imbalance, and systemic immune dysregulation, which in turn will negatively affect nutrient uptake and utilization, growth, and survival rate, are also frequently observed. Modern broiler hybrids seem to be particularly susceptible to heat stress, since the high body heat resulting from their great metabolic activity may exacerbate this phenomenon [23]. Metabolic disorders resulting from other microclimate alterations (such as cold, hypoxia, and light/dark hour changes) are less frequent and quite nonspecific [22].

Management defects related to *housing conditions* and *social interactions* among animals are strictly related to each other. One of the most frequent welfare problems in broiler chickens is **contact dermatitis** (i.e., hock burns, breast burns, and foot pad dermatitis), which is caused by continuing contact and pressure of the skin of the breast, hocks, and feet against humid and soiled bedding. In particular, **footpad dermatitis** (FPD) has the greatest relevance (**Figure 4B**). It is also known as pododermatitis and represents a condition that is characterized by inflammation and necrotic lesions, ranging from superficial to deep on the plantar surface of the footpads and toes. Deep ulcers may also lead to abscesses and thickening of underlying tissues and structures. Several environmental factors such as litter material, moisture depth and amendments, drinker design and management, and stocking density may influence FPD development. Indeed, a straw, wet, thin, and acidifier-added litter and small drinker cups and higher stocking densities have been reported to be associated with a greater incidence of FPD [24]. **Feather-pecking**, which is defined as a nonaggressive behavior whereby birds peck at and/or pull out the feathers of conspecifics, represents one of the most significant welfare concerns in laying hens resulting in feather damage, feather loss, wounds, pain, cannibalistic pecking, and death. Development of feather-pecking has been associated with different causative factors, one of the most important being the inhibition of foraging behaviors (such as ground-pecking or lack of environmental stimuli) and lack of early life access to litter [25].

The selection procedures focused on a high growth rate may cause specific diseases of the energy-supplying organs (in particular the intestine and the liver), as a result of the developing imbalances between oxygen supply and oxygen requirement. In particular, **fatty liver-hemorrhage syndrome** (FLHS) is frequently observed in laying hens, while broiler chicken gut may show malabsorption syndrome [26]. High growth rates, as well as high body weights and low levels of activity, are also frequently associated with the development of lameness of various degrees of severity. It is most prominent in rapidly growing males, with **leg deformities** such as angular bone deformity (valgus-varus), dyschondroplasia, and spondylolisthesis (kinky back) accounting for 65–80% of the noninfectious causes of lameness in broiler chickens. Modern fast-growing strains may also present an increase in skeletal muscle myopathies, such as **white striping** and **wooden breast**. In turkeys, focal avascular or ischaemic necrosis (osteochondrosis) of articular cartilage, avulsion fractures and ligament damage at the intertarsal joint or femorotibial joints, and spontaneous fracture of the femur may also occur [21]. Finally, **pulmonary arterial hypertension** (PAH, also known as ascites syndrome and pulmonary hypertension syndrome) is one of the most common diseases observed worldwide in fast growing broilers (**Figure 4C**). This disease can be attributed to the fast growth-related imbalances between cardiac output and the anatomical capacity of the pulmonary vasculature to accommodate ever-increasing rates of blood flow, as well as to an inappropriate degree of constriction maintained by the pulmonary arterioles. Other common **cardiovascular diseases** associated with rapid growth are the sudden death syndrome (SDS) in broilers and hypertrophic cardiomyopathy (HCM), spontaneous turkey cardiomyopathy (STC), and aortic dissecting aneurysm in turkeys [22].

2.5 Fish

Fish class is the biggest and the most differentiated among vertebrates. Fishes are adapted to different extreme situations as their evolutional success depends on their ability to thrive in a variable medium: water. One of the most remarkable examples of the strict connection between fishes and water is the fishes' inability to regulate their internal temperature (ectothermic animals). Nowadays aquaculture is one of the more sustainable and economically favorable sources of animal protein. Studies on the effect and pathological results of the fishes' "disturbed habitat" are well known due to common compromised (naive) situations. Considering wild habitats, we must sentence that they are strongly impacted by human activity (pollution, overfishing, and introduction of non-indigenous organisms), and this makes it difficult to define what is to be considered normal, not normal, or sub-normal for fishes in a specific situation. In farmed animals, the effect of "disturbed habitat" can sometimes become more evident and dramatic than into the wild [27]. Moreover, the severity of a given disease is dependent on the intricate interaction of numerous variables of the host, the pathogen, and the environment, among which the environment is the less-known factor [28]. In addition, early signs of suffering in fishes are difficult to relate to a specific disease by inexperienced staff. Commonly, acute stressed fishes show color changes because the melanin pigmentation in skin is under neuroendocrine control in fish and it is thus affected by hormones such as epinephrine involved in the first step of stress reaction. If a stress factor persists for a longer time, other hormones, such as cortisol, become dominant (chronic stress). The chronic stress induces immunodeficiency that causes a higher incidence of opportunistic disease outbreaks. Despite the difficulties explained above, in the following section, we will focus on the main stress factors that can impact fish welfare.

Among *abiotic ambient factors,* there are all the physical and chemical water parameters such as temperature, conductivity, salinity, turbidity, hardness, dissolved oxygen and other gasses, pH, ammonia and other nitrogen compounds, metals, pesticides, etc. Fishes can handle an open range of variations for each parameter without showing recognizable signs of disease or suffering, thus accumulating chronic stress. Out of these ranges, water quality parameters can influence acute stress along with **high mortality** showing or not respiratory symptoms. More frequently, considering the synergistic effect of water-quality parameters, only subclinical evidences like a **reduction of productivity and reproduction, dissimilarity of age classes** (for wild stocks), or a higher impact of some **infectious agents or tumors** (if a carcinogenic pollutant is suspected) can be noted [29]. Focusing on the farm self-pollution, due to organic wastewater and nitrogen compound discharge (e.g., ammonia), a direct **damage at the fish gills** (acute gill disease) is evident due to a decreased dissolved oxygen. This acute gill disease is easily detected in fishes with an acute respiratory distress shown by a higher frequency of gill opercula movements. On the contrary, in case of prolonged mild problems, fishes develop a "chronic gill disease" characterized by a fusion of secondary lamellae [30] and a typical fish silhouette called "snake head shape" due to the contrast between a large and triangular head, deformed for the enlarged opercula, and a thin body. In fact, the low level of oxygen blood saturation causes a growth failure for the inability to optimally metabolize food nutrients.

The *housing system* in aquaculture management must take into consideration the different biology, ecology, and natural behaviors of individual fish species. The space, design, composition material for tanks, pools, basins and nets, water source, flow and change, lighting and photoperiod, etc. must be taken into account. An inappropriate housing system determines lower growth performances and a higher incidence of **opportunistic diseases** due to chronic stress [31]. The *degree of social interaction* among fishes, with the main critical point of the animal density, is different in extensive farming when compared to the intensive one: the first is closer to a wild condition while the second is richer of health-limiting conditions. In nature, high animal density happens only for short times (i.e., migration for the reproduction or for feed) but in farmed fishes is a constant need. Over-density causes **traumatic lesions by bite** (skin erosion, ulcers, and body mutilation) (**Figure 4D**) and fast deterioration of water-quality parameters. Similar consequences can be observed as a result of *husbandry practices*, such as selection, artificial reproduction, handling, transport, and net confinement, especially if carried out without suitable tools or by unskilled workers (**Figure 4E**). **Infections** caused by opportunistic bacteria or fungi (Oomycetes) such as *Flavobacterium* spp., *Columnaris* disease (fin or skin rot), or *Saprolegnia* spp. (water mold infection) can develop into skin or gill injuries (**Figure 4F**). Sometimes, if fish density is high and the water quality and exchange low, **parasites** such as barnacles or motile ciliates can also provoke a massive outbreak with evident skin hemorrhages and erosions. At the same time, also common aquatic bacteria such as motile *Aeromonas* spp., *Pseudomonas* spp., or *Vibrio* spp. can cause septicaemia characterized by skin, gills, and internal organ hemorrhages, pop eyes, and skin ulcerations. Regarding feeding, as fish are ectothermic, periods of food deprivation may be less detrimental than in endotherms. For this reason, temporary starvation prior to transport, treatment of disease, or any other kind of handling procedures is highly recommended to reduce physiological stress [27]. However, an inappropriate food composition or feeding procedure can generate **gut problems** like enteritis and **size inhomogeneity** in the fish stock [32].

3. Zoo and wild animals

The literature merges concepts and definitions of modern zoos and aquaria as "centers for education and conservation." In this sense, their animals are considered as "ambassador guests" or even stakeholders of the zoological institutions. Husbandry procedures impact directly on the welfare of every managed animal species and should be carried out with "human care" regardless of the context, artificial, semi-natural, or wild; this highlights the importance of precise estimation of the resources provided to the animals, by humans.

But, is this always the case? Advocacy groups often claim that animals would only live a good life "worth living," if they were left into the wild. In reply to this controversy stands the extremely relevant anthropic detrimental influence on the environment and a precise definition of "ex situ" as: "conditions under which individuals are spatially restricted with respect to their natural spatial patterns or those of their progeny, are removed from many of their natural ecological processes and are managed on some level by humans" [33].

This concept has created a lot of confusion in the years, on whether the artificial environment could really provide excellent welfare conditions, to be evaluated directly on each individual. Welfare indicators in fact, vary among and within species, and depend directly on human interests and uses of them. This is particularly true for nondomestic species where individual case relevance is rare, fragmented, and often requiring comparison with their wild counterpart.

For this reason, modern zoological institutions tend to mimic the irreplaceable wildlife observations and provide the animals with environmental resources extrapolated from previous *ex situ* experiences and consolidated in best practice guidelines [34]. Careful attention is paid to *animal management processes*, starting from animal acquisition and transport, quarantine and acclimation, and introduction into social group. Exhibits with multiple species must take into consideration social compatibility, both intra and interspecific. The density and distribution of most species should be compatible with the space provided, allowing the expression of natural behaviors and guaranteeing individual safety, thus avoiding **undesired dominance and aggression** [35].

Zoo and aquarium "artificial habitat" construction takes into consideration the preparation of controlled microclimate systems [35]. For example, indoor air temperature, ventilation, and filtered aeration prevent transmission of **respiratory pathogens** (*Aspergillus* spp., *Mycobacterium* spp., etc.) in most air breathing species. Appropriate lighting and photoperiod allow a natural circadian rhythm to regulate hormonal cycles, reproduction, and molt in most species [36]. For aquatic and semi-aquatic species, the installation and management of "Life Support System" provides computerized systems to control filtration, disinfection, and chemical/physical water parameters suitable for the species maintained. Mechanical filtration removes particulate solid matters and the complex system of biological filtration avoids direct contact with all the toxic compounds originating from nitrogen life cycle. Water temperature control is mandatory since it could literally limit survival of species originating from different climates. Inappropriate levels of pH, salinity, and hardness might lead to **chronic stress** or even death. Water disinfection and oxidation must be under strict control to avoid damages or losses from accidental increased redox potential, lethal for fish and invertebrates, and seriously **damaging skin and eyes** of aquatic reptiles and mammals [37].

Health affects the animal's welfare and the quality of its life. Veterinary programs address general and specific issues such as nutrition, reproduction, and management of geriatric individuals. *Unbalanced diets* can lead, for example, to **abnormal growth, gout,** or **hypovitaminosis**, to even impossibility to thrive [35]. Poor fitness affects breeding and lifespan, but also physical appearance and behavior of the individuals, influencing the human perception of their role for conservation or "ambassadors of the species." Zoos and aquaria maintain animals in good physical, social, and mental health, to fulfill their mission and promote *ex situ* conservation.

Wildlife is also strongly influenced by human impact on the environment, only from a different perspective. The correlation between "disturbed habitat" and pathology is in fact not always clear, nor evident for wild animals: disease and death are in fact processes of the normal circle of life, and can be considered as unnatural problems only when caused by human interference. As stated by the World Association of Zoo and Aquaria in Caring for wildlife (2015) [38]: "we affect animals by destroying their habitats, polluting their environment, introducing invasive species into their ecological systems, building structures in flight-paths, tilling the land, cutting trees, driving cars, burning fuel, and on and on...."

An increased food demand, an intensification, and mechanization in agriculture, including use of chemical products, led to a widespread decline in farmland biodiversity and remarkable change of landscapes and habitats. The use of pesticides facilitates farmers' work, thus menaces the environment and its living creatures. In the European rice fields, the butterfly *Lycaena dispar*, an important environmental indicator, is in decline due to the massive use of herbicides; and similar events occurred in Japan as well [39]. "The European Red List of Saproxylic Beetles" (2018) [40] highlights the importance of these beetles in the forest ecosystems and their dependence on dead and decaying wood. They are involved in numerous processes but often ruined by the wrong perception that deadwood is a sign of neglected forest management.

The effects of climate warming are recognized by everyone and lead to desertification in many countries, provoke unprecedented disastrous events, and affect ecosystems and species survival around the world, including our own. Glaciers melting at an increasing speed directly affect polar bears (*Ursus maritimus*) and Arctic environment, leading to disappearance of their habitat and food resources.

Invasive species became a very relevant problem [41]: the gray squirrel (*Sciurus carolinensis*), introduced in the last century in Europe, is a threat to the native red squirrel (*S. vulgaris*). Gray squirrels compete for resources and, in Britain, are a reservoir for a virus highly pathogenic for red squirrel, inducing a disease-mediated competition between the species. Other countries could face the same problem in future: in Japan, *S. vulgaris* is a common pet with high risk of uncontrolled release and impact on the native *S. lis* [42]. The introduction of the popular American pet red-eared slider caused similar threats imposing major conservation activities to preserve the native European pond turtle [43].

Artificial lights can affect plant's photosynthesis, circadian rhythm, visual perception, and spatial orientation and can disorient many nocturnal species. Preys around lamps attract bats that also become more detectable by birds of prey. New road constructions are welcomed with enthusiasm, but they fragment habitats and represent insuperable dangerous barriers for crossing animals.

Coexistence between domestic and wild species can spread **transmissible diseases**. Infectious keratoconjunctivitis originates from infected livestock and passes to alpine chamois (*Rupicapra rupicapra*) and alpine ibex (*Capra ibex*) that graze close to each other. Another type of challenge is represented by the return of wolves in the Alps, since farmers do not tolerate their predation on livestock. The authorities promote preventive measures and compensation for the damage, but the conflict is strongly due.

Overexploitation of natural resources is a worldwide-recognized problem and animal collectors have a huge impact on biodiversity. Oriental medicine utilizes parts of wild animals (tigers, rhinos, sharks, seahorses, etc.). The indiscriminate fishing of the totoaba, a very popular fish, whose swim bladder represents an unremarkable black market value, has brought the small cetacean vaquita on the verge of extinction.

Finally, in the last decades, humans developed new types of sports and touristic attractions with animal direct contact and experience. As an example, ski mountaineering, a dangerous and exciting popular sport in the Alps, can be lethal for alpine animals, such as the black grouse (*Tetrao tetrix*) or rock ptarmigan (*Lagopus muta*). Photo safaris and the desire to "get closer" to wilderness, at all costs, are a growing "leisure product" which is very disgraceful to nature.

4. Conclusion

In 1915, Cannon defined stress as a "perturbation of the homeostasis, the coordinated physiological process which maintains a steady state in the organism" [44]. A persistent stress condition may result in psychological and physiological pathology.

DOI: http://dx.doi.org/10.5772/intechopen.84872

From the perspective of domestic/wild animals, these pathologies may occur at a clinical or subclinical level and may manifest as altered behavior, decreased immune protection that impacts disease susceptibility, or altered metabolism that impacts either growth, production, or a combination of these responses.

This chapter represents a non-exhaustive list including only a few examples of scientific evidence that farm animals or wildlife face multiple threats related to stressing factors/habitat disturbance, by direct or indirect human impact on the environment. While it is clear that we have been the major cause of these dramatic changes, we are also growing a generalized protective conscience towards natural resources. A virtuous search for new technologies and alternative human behavioral changes is now mandatory to minimize our impact and foster our survival on this planet [45, 46].

In this contest, animal welfare relates to more than merely the physical health of an individual. Animal welfare means how an animal is coping with the conditions in which it lives [47]. An animal is in a good state of welfare if it is in an appropriate social context, healthy, comfortable, well nourished, safe, able to express innate behavior, and if it is not suffering from unpleasant states such as pain, fear, and distress. Good animal welfare requires disease prevention and veterinary treatment, appropriate shelter, management, nutrition, humane handling and, ultimately, humane slaughter. Animal welfare refers to the state of the animal and protecting an animal's welfare means providing for its physical and mental needs [8].

In recent years, a great and growing attention is paid to the aspects of health and welfare of reared animal species. For farm animals, the initiative of people to care about the welfare of farm animals is based on their moral attitude and concern for the right and wrong treatment of animals, with presumed opposition to overexploitation and/or cruelty towards animals [48]. There is also growing concern for many consumers in Europe about farm animal welfare since it is becoming increasingly recognized as an important attribute of food safety and quality [48, 49]. To enhance animal welfare, a first essential step is to help animals to cope with their environment. Two different approaches can be used: firstly we propose to adapt the environment to the animals by improving management practices and housing conditions. This approach requires the active involvement of all stakeholders: veterinarians, behaviorists, animal scientists, the industrial farming sector, the food processing and supply chain, and consumers of animal-derived products. Secondly, we can create rearing conditions that better "prepare" animals for the environment in which they will be kept in later stages of their life.

Also for zoo animals, as habitats and ecosystems become increasingly altered and populations evermore impacted by human activities, a growing number of species will require some form of management of both individuals and populations to ensure their survival. Zoo and aquaria aim to fulfill this role. Whereas zoos and aquariums of the past were places where animals were "displayed" for the pleasure of visitors, today's zoos and aquariums are centers for conservation. They must ensure that the conditions for animals in their care are the best that can be delivered, providing environments that focus on the animals' physical and behavioral needs. To accomplish this, it is necessary to make sure that animal care staff have relevant scientific training and expertise, developing and maintaining a staff culture that practices regular reporting and monitoring of animals' behavior and health, employing veterinarians, biologists, welfare scientists, and behavioral experts, introducing different enrichments that provide challenges, choices, and comfort to animals to maximize their psychological health. In fact, the major features of animal welfare that are relevant to zoos and aquariums merge the following aspects: meeting animals' basic survival needs for food, shelter, health, and safety and enhance their welfare above this survival minimum by increasing opportunities for animals to have positive experiences, focused, for example, on their comfort, pleasure, interest, and confidence. Although visitors' direct impact on animal welfare is limited, their expectations have risen sharply and support the zoo and aquarium commitment to keep animals healthy and engaged, by also utilizing environmental enrichment skills [38].

To conclude, considering the enormous number of animals whose life conditions are affected by human habits, all possible efforts can and must be made to improve their status in order to ban welfare-compromising procedures and practices as soon as possible.

Conflict of interest

The authors declare no potential conflicts of interest with respect to the research, authorship, and/or publication of this article.

Author details

Maria Teresa Capucchio[1*], Elena Colombino[2], Martina Tarantola[1], Davide Biagini[2], Loris Giovanni Alborali[3], Antonio Marco Maisano[3], Federico Scali[3], Federica Raspa[1], Emanuela Valle[1], Ilaria Biasato[2], Achille Schiavone[1], Cristian Salogni[3], Valentina Bar[1], Claudia Gili[4] and Franco Guarda[1]

1 Department of Veterinary Sciences, University of Torino, Italy

2 Department of Agricultural, Forest and Food Sciences, University of Torino, Italy

3 Istituto Zooprofilattico Sperimentale di Lombardia e Emilia Romagna, Brescia, Italy

4 Department of Science and Veterinary Services at Acquario di Genova, Costa Edutainment S.p.A., Genova, Italy

*Address all correspondence to: mariateresa.capucchio@unito.it

References

[1] Van Andel J, Bakker JP, Snaydon RW, editors. Disturbance in Grasslands. Causes, Effects and Processes. Dordrecht: Junk, Publ.; 1987. 316 p

[2] European Council. Treaty of Lisbon amending the Treaty on European Union and the treaty establishing the European Community (2007/C 306/01). 2007. Available from: http://eur-lex.europa.eu/legal-content/EN/TXT/HTML/?uri=CELEX:12007L/TXT&from=EN [Accessed: Feb 25, 2019]

[3] OIE. Introduction to the Recommendations for Animal Welfare, Chapter 7.1.1. Terrestrial Animal Health Code. 2017. Available from http://www.oie.int/index.php?id=169&L=0&htmfile=chapitre_aw_introduction.htm [Accessed: Feb 25, 2019]

[4] Mellor DJ. Updating animal welfare thinking: Moving beyond the "Five freedoms" towards "A life worth living". Animals. 2016;**6**(3):21. DOI: 10.3390/ani6030021

[5] Nordquist RE, van der Staay FJ, van Eerdenburg FJCM, Velkers FC, Fijn L, Arndt SS. Mutilating procedures, management practices, and housing conditions that may affect the welfare of farm animals: Implications for welfare research. Animals. 2017;7(2):12. DOI: 10.3390/ani7020012

[6] EFSA Panel on Animal Health and Welfare (AHAW). Scientific Opinion on the welfare of cattle kept for beef production and the welfare in intensive calf farming systems. EFSA Journal. 2012;**10**(5):2669. 166 p. DOI: 10.2903/j.efsa.2012.2669

[7] Tarantola M, Valle E, De Marco M, Bergagna S, Dezzutto D, Gennero MS, et al. Effects of abrupt housing changes on the welfare of Piedmontese cows. Italian Journal of Animal Science. 2016;**15**:103-109. DOI: 10.1080/1828051X.2015.1128691

[8] Animal Welfare Institute. The Critical Relationship Between Farm Animal Health and Welfare. 2018. Available from: https://awionline.org/sites/default/files/uploads/documents/FA-AWI-Animal-Health-Welfare-Report-04022018.pdf [Accessed: Dec 7, 2018]

[9] Biasato I, Biasibetti E, Biagini D, Bruatto G, Cenacchi G, Guarda F, et al. Spontaneously occurring intramural coronary arteriosclerosis in regularly slaughtered veal calves and beef cattle: A screening study about prevalence and histopathological features. Journal of Veterinary Cardiolology. 2018;**20**:55-63. DOI: 10.1016/j.jvc.2017.12.001

[10] Caroprese M, Casamassima D, Rassu SPG, Napolitano F, Sevi A. Monitoring the on-farm welfare of sheep and goats. Italian Journal of Animal Science. 2009;**8**:343-354

[11] EFSA Panel on Animal Health and Welfare (AHAW). Statement on the use of animal-based measures to assess the welfare of animals. EFSA Journal. 2012;**10**(6):2767. 29 p. DOI: 10.2903/j.efsa.2012.2767

[12] Welfare Quality®. Welfare Quality® assessment protocol for pigs (sows and piglets, growing and finishing pigs). In: Welfare Quality® Consortium, Lelystad, Netherlands. 2009. pp. 1-122

[13] EFSA AHAW Panel (EFSA Panel on Animal Health and Welfare). Scientific Opinion concerning a multifactorial approach on the use of animal and nonanimal-based measures to assess the welfare of pigs. EFSA Journal. 2014;**12**(5):3702. 101 p. DOI: 10.2903/j.efsa.2014.3702

[14] Bolhuis JE, Schouten WGP, SChrama JW, Wiegant VM. Behavioural

development of pigs with different coping characteristics in barren and substrate enriched housing conditions. Applied Animal Behaviour Science. 2005;**93**:213-228. DOI: 10.1016/j.applanim.2005.01.006

[15] Thomson J, Friendship R. Digestive system. In: Zimmerman JJ, Karriker LA, Ramirez A, Schwartz KJ, Stevenson GW, editors. Diseases of Swine. 10th ed. Chicester: Wiley; 2012. pp. 208-211

[16] Brambilla G, Civitareale C, Ballerini A, Fiori M, Amadori M, Archetti LI, et al. Response to oxidative stress as a welfare parameter in swine. Redox Report. 2002;**7**(3):159-163. DOI: 10.1179/135100002125000406

[17] Ivester KM, Couëtil LL, Zimmerman NJ. Investigating the link between particulate exposure and airway inflammation in the horse. Journal of Veterinary Internat Medicine. 2014;**28**:1653-1665. DOI: 10.1111/jvim.12458

[18] Henderson AJZ. Don't fence me in: Managing psychological well being for elite performance horses. Journal of Applied Animal Welfare Science. 2007;**10**(4):309-329. DOI: 10.1080/10888700701555576

[19] Casey RA. Clinical problems associated with the intensive management of performance horses. In: Waran N, editor. The Welfare of Horses. Springer: Dordrecht; 2007. pp. 19-44. DOI: 10.1007/978-0-306-48215-1_2

[20] Christensen JW, Sondergaard E, Thodberg K, Halekoh U. Effects of repeated regrouping on horse behaviour and injuries. Applied Animal Behaviour Science. 2011;**133**(3-4):199-206. DOI: 10.1016/j.applanim.2011.05.013

[21] Pollitt CC, Visser MB. Carbohydrate alimentary overload laminitis. Veterinary Clinics of North America: Equine Practice. 2010;**26**(1):65-78. DOI: 10.1016/j.cveq.2010.01.006

[22] Julian RJ. Production and growth related disorders and other metabolic diseases of poultry—A review. Veterinary Journal. 2005;**169**(3):350-369. DOI: 10.1016/j.tvjl.2004.04.015

[23] Tsiouris V, Georgopoulou I, Batzios C, Pappaioannou N, Ducatelle R, Fortomaris P. Heat stress as a predisposing factor for necrotic enteritis in broiler chicks. Avian Pathology. 2018;**47**(6):616-624. DOI: 10.1080/03079457.2018.1524574

[24] Shepherd EM, Fairchild BD. Footpad dermatitis in poultry. Poultry Science. 2010;**89**(10):2043-2051. DOI: 10.3382/ps.2010-00770

[25] Hartcher KM, Tran KT, Wilkinson SJ, Hemsworth PH, Thomson PC, Cronin GM. The effects of environmental enrichment and beak-trimming during the rearing period on subsequent feather damage due to feather-pecking in laying hens. Poultry Science. 2015;**94**(5):852-859. DOI: 10.3382/ps/pev061

[26] Scheele CW. Pathological changes in metabolism of poultry related to increasing production levels. Veterinary Quarterly. 1997;**19**(3):127-130. DOI: 10.1080/01652176.1997.9694756

[27] Ashley PJ. Fish welfare: Current issues in aquaculture. Applied Animal Behaviour Science. 2007;**104**(3-4):199-235. DOI: 10.1016/j.applanim.2006.09.001

[28] Hedrick RP. Relationships of the host, pathogen, and environment: Implications for diseases of cultured and wild fish populations. Journal of Aquatic Animal Health. 1998;**10**(2):107-111. DOI: 10.1577/1548-8667(1998)010<0107:ROTHPA>2.0.CO;2

[29] Svobodova Z, Kolarova J. A review of the diseases and contaminant related mortalities of tench (*Tinca tinca L.*). Veterinární Medicína. 2004;**49**(1):19-34

[30] Strzyzewska E, Szarek J, Babinska I. Morphologic evaluation of the gills as a tool in the diagnostics of pathological conditions in fish and pollution in the aquatic environment: A review. Veterinární Medicína. 2016;**61**(3):123-132

[31] Gebregeziabhear E, Ameha N. The effect of stress on productivity of animals: A review. Journal of Biology, Agriculture and Healthcare. 2015;**5**(3):165-172

[32] Booman M, Forster I, Vederas JC, Groman DB, Jones SRM. Soybean meal-induced enteritis in Atlantic salmon (*Salmo salar*) and Chinook salmon (*Oncorhynchus tshawytscha*) but not in pink salmon (*O. gorbuscha*). Aquaculture. 2018;**483**:238-243. DOI: 10.1016/j.aquaculture.2017.10.025

[33] IUCN/SSC. Guidelines on the Use of Ex Situ Management for Species Conservation. Version 2.0. Gland. Switzerland: IUCN Species Survival Commission. 2014

[34] Miller RE, Lamberski N, Calle P, editors. Miller-Fowler's Zoo and Wild Animal Medicine Current Therapy. Vol. 9. St. Louis, Missouri: Elsevier Saunders; 2018. 768 p

[35] Fowler ME, Miller RE, editors. Fowler's Zoo and Wild Animal Medicine. Vol. 8. St. Louis, Missouri: Elsevier Saunders; 2014. 792 p

[36] Mo G, Gili C, Ferrando P. Do photoperiod and temperature influence the molt cycle of *Phoca vitulina* in captivity? Marine Mammal Science. 2000;**16**(3):570-577. DOI: 10.1111/j.1748-7692.2000.tb00952.x

[37] Gulland FMD, Dierauf LA, Whitman KL, editors. CRC Handbook of Marine Mammal Medicine. 3rd ed. Boca Raton: CRC Press, Taylor & Francis Group; 2018. 1124 p

[38] Mellor DJ, Hunt S, Gusset M, editors. Caring for Wildlife: The World Zoo and Aquarium Animal Welfare Strategy. Gland: WAZA Executive Office; 2015. 87 p

[39] Katayama N, Baba YG, Kusumoto Y, Tanaka K. A review of post-war changes in rice farming and biodiversity in Japan. Agricultural Systems. 2015;**132**(2015):73-84. DOI: 10.1016/j.agsy.2014.09.001

[40] Cálix M, Alexander KNA, Nieto A, Dodelin B, Soldati F, Telnov D, et al. European Red List of Saproxylic Beetles. Brussels, Belgium: IUCN; 2018

[41] Global Invasive Species Database. Available from: http://www.iucngisd.org/gisd/ [Accessed: Dec 18, 2018]

[42] Shar S, Lkhagvasuren D, Bertolino S, Henttonen H, Kryštufek B, Meinig H. Sciurus vulgaris. The IUCN Red List of Threatened Species. 2016. DOI: 10.2305/IUCN.UK.2016-3.RLTS.T20025A22245887.en

[43] Ottonello D, Sebastiano S, Jesu R, Gili C, Genta P, Canessa S, et al. Conservation of the European pond turtle through population reinforcement in Liguria, Northern Italy. In: Soorae PS, editor. Global Re-introduction Perspectives: 2016. Casestudies from Around the Globe. Gland, Switzerland: IUCN/SSC Reintroduction Specialist Group and Abu Dhabi, UAE: Environment AgencyAbu Dhabi; 2016. pp. 62-66

[44] Cannon WB, editors. Bodily Changes in Pain, Hunger, Fear, and Rage: An Account of Recent Researches into the Function of Emotional

Excitement. New York: Appleton and Company; 1915. 334 p

[45] Francesco P. Laudato si'-Lettera enciclica sulla cura della casa comune. Cinisello Balsamo: Edizioni San Paolo srl; 2015. 212 p

[46] Sustainable Development Goals. 2015. Available from: https://www.un.org/sustainabledevelopment/sustainable-development-goals/ [Accessed: Jan 2, 2019]

[47] Broom DM. Applied the scientific assessment of animal welfare. Animal Behaviour Science. 1988;**20**(1-2):5-19

[48] Ostojić DA, Aleksić D, Petrović MM, Pantelić V, Stanišić N, Petrović VC, et al. Beef cattle welfare—Risks and assurance. Biotechnology in Animal Husbandry. 2015;**31**(3):313-326

[49] Blokhuis H, Keeling LJ, Gavinelli A, Serratosa J. Animal welfare's impact on the food chain. Trends Food Science and Technology. 2008;**19**:79-87